CHANCE VOUGHT F7U-1 CUTLASS BY TOMMY H. THOMASON

Above, but for a series of problems and unfortunate events, this encounter between a Russian MiG-15 and the F7U-1 Cutlass might have occurred in the skies above Korea in 1951. (Author's collection)

INTRODUCTION

For the aviation enthusiast, one of the iconic images of the Korean War is the air-to-air dogfight between the Russian MiG-15 and the U.S. Air Force F-86 Sabre. Unlike most fighters at the time, both had swept-wings. This feature, based on WWII German research, increased their speed in level flight and dives compared to jet fighters with straight wings like the Air Force F-80s and F-84s, and the Navy F9Fs and F2Hs. The slower jets were therefore mostly relegated to bombing and close-air support missions, augmenting the propeller-driven fighters that they had replaced. The Navy was developing a swept-wing fighter, the Chance Vought F7U-1 Cutlass — also named after a bladed weapon like the Sabre — but its deployment had been delayed by a series of problems and misfortunate events. As a result, the Sabre pilots didn't have to share the glory of their MiG kill-ratio supremacy.

Not deploying the Cutlass in the early 1950s wasn't for lack of a strong Navy advocate or a valiant effort by Vought. In 1945, CDR A.B. Metsger was the newly-assigned head of the fighter design branch (familiarly referred to as the fighter class desk) at the Navy's Bureau of Aeronautics (BuAer). He strongly believed that the Navy needed to transition to swept-wing jet fighters as soon as possible. It was not a widely shared vision at the Bureau. The concern was the ability to operate safely from aircraft carriers. The jet engine's shortcomings — lack of thrust at low speeds and prodigious consumption of fuel at any speed — were bad enough. The high stall speed and poor handling qualities of the swept wing at low speeds were an even greater impediment to acceptance. Most therefore believed that a combination of the propeller and the jet engine was necessary, either a compound arrangement (both a piston engine and a jet engine in the same aircraft) or the newly-conceived turboprop engine, in which a turbine in the exhaust of a jet engine turned the propeller via a shaft going forward to a large rpm-reduction gearbox.

At the time, engineers were also dealing with a newly-discovered aerodynamic phenomenon, compressibility, discovered by pilots diving propeller-driven fighters at high speed. The encounters were sometimes frightening and occasionally fatal. The change in handling qualities of airplanes at transonic speeds was not well understood. It was like going over Niagara Falls in a barrel although the survival rate was much better. The airframe buffet, wing drop, and aileron buzz were attention-getting. The real problem was that the flight controls became less effective, or in the case of pitch, seemingly reversed in function. What was worse, the jet pilot could get even deeper and quicker into transonic trouble because there was no propeller drag limiting the Mach number attainable and facilitating recovery.

Delaying the onset of compressibility to higher airspeeds (actually, Mach numbers) also increased acceleration as well as maximum speed in a dive. This was a particularly significant advantage if a pilot was trying to lose someone who was on his tail or stay with someone who he was trying to shoot down. Increasing the critical Mach number also expanded the "buffet boundary" that affected maneuverability at high speeds and altitudes. Fighters with low buffet boundaries were at the mercy of fighters with higher buffet boundaries if a turning fight developed.

Both the Army and the Navy had initiated jet engine and jet fighter programs early in World War II. However, these were conventional in basic planform, with straight wings.

While faster than propeller-driven airplanes with reciprocating engines, they were limited in acceleration and speed in a dive by transonic drag and handling quality problems. Following the war, the allies (which then included Russia) became aware of German swept-wing research. It was already becoming evident that thinner airfoils, relative to their chord, increased the critical Mach number. Unfortunately, thin wings had significant drawbacks, among them volume for fuel and adequate aeroelastic stiffness. The swept wing provided a geometric way to thin the airfoil aerodynamically while retaining a reasonable thickness, raising achievable Mach numbers in level flight on the same thrust from about 0.8 to 0.9, or about 70 mph (61 knots) at 30,000 feet.

Both the Army and the Navy had ordered similar straight-wing jet fighters from North American in the last

Top-to bottom: The Lockheed P-80 was also pressed into service in the Korean War as a bomber. In this case, it must have not been going far, as the tip tanks have been removed so two more bombs can be carried on the wingtip stores stations. (USAF) Like the P-80, another early Air Force jet fighter, the Republic F-84, was pressed into service for strike missions. While its higher speed made it more survivable than propeller-driven airplanes in that role, it was too slow to be an effective fighter when the Russian MiG-15s arrived. (USAF) Although Grumman's F9F Panther was initially designed without any external stores capability, a modification was quickly developed to enable it to carry bombs and rockets. Panthers with this capability were initially designated F9F-2B, with the B standing for an armament modification. This F9F-2B, armed with small bombs and antitank HVARs, is being prepared for launch from Princeton on 20 April 1951. (NA 80-G-428389) The McDonnell F2H-2 Banshee was also modified to carry external stores. One set of armament pylons was located well out on the wing to reduce loads on the wing structure when maneuvering with the increased weight. Here, bombs are being loaded and armed prior to a mission from Essex during the Korean War. (NA 80-G-432724)

year of the war, the XP-86 and the XFJ-1 respectively. In August 1945, however, the Army directed North American to evaluate a swept-wing configuration based on the performance potential predicted by German wind tunnel testing. In November, based on the results of the evaluation, the Army directed North American to redesign the XP-86 with a swept wing. Opting for a higher top speed that required a higher angle of attack for the same lift and reduced the maximum lift coefficient was easy for the Air Force. It wasn't for the Navy, which chose to retain the

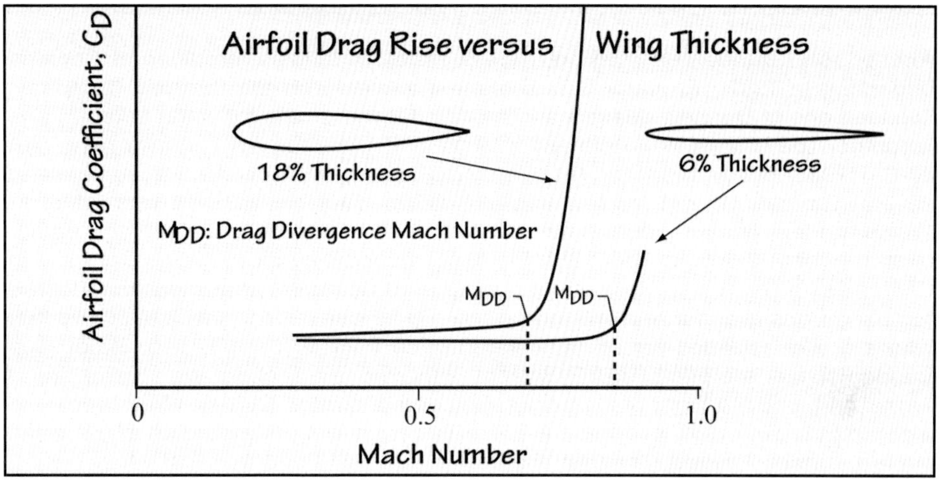

Top-to-bottom: The Ryan XF2R-1 combined a 1700-hp General Electric TG-100 turboprop engine in the nose with a 3,000-lb thrust Westinghouse J34 jet engine in the tail. It was almost as fast as the early jets and had far superior takeoff, climb, and wave-off performance. The propeller limited its speed potential, however, particularly in a dive whereas the jet fighter was already demonstrating near-sonic performance in a dive, which was indicative of its much better top-speed potential. (NA 80-G-703586) Wind tunnel testing demonstrated that thinner wings were necessary to delay the onset of rapidly increasing drag with Mach number. The drag rise is even more significant than shown, because it is proportional to the drag coefficient multiplied by the square of the velocity. (Author) As part of the Navy's program to transition to the jet age as quickly as possible, North American received a contract in January 1945 for the FJ-1 jet fighter powered by a single General Electric TG-180 engine. Like the Navy's other jet airplanes at that time, it did not have an ejection seat or cockpit pressurization. (USN) Based on German wind tunnel test data, the Army Air Forces directed North American to incorporate swept wings on its counterpart to the Navy's FJ-1. The XP-86's first flight was on 1 October 1947 with a 4,000-lb thrust Chevrolet-built J35-C-3 instead of the 5,200-lb thrust General Electric J47-GE-13 that would be installed for qualification and production. Development was relatively trouble free, with squadron deliveries beginning in 1949. The gross weight was about 16,000 lbs. (USAF)

Above, John Dunne designed and flew a series of tailless airplanes in England beginning in 1907. An early U.S. airplane manufacturer, the Burgess Company, licensed one of Dunne's designs and modified it with floats for water takeoffs and landings. The U.S. Navy purchased several in 1914 as the AH-7 shown here. (NA 80-G-417255) Below, following the victory in Europe, the Army Air Forces obtained several German aircraft for test and evaluation. This Me 163B-1a was taken to Freeman Field, Indiana, during the summer of 1945 and assigned foreign equipment number FE-500. In April 1946, the Army moved it by air to Muroc Dry Lake in California for flight test. Because of operation of its rocket motor was considered too dangerous even for research purposes, a B-29 was used to tow it to 30,000 feet where it was released for an unpowered glide back to the lakebed. Major Gus Lundquist was the pilot. (USAF) Bottom, after two years of design studies, wind tunnel test, and the flight of a configuration demonstrator, Curtiss-Wright received a contract in July 1942 for three XP-55s with a large wing mounted aft and a small canard surface forward. The first one flew in July 1943 and was lost when it went out of control in November. Disappointing performance and handling quality issues precluded production. (USAF)

straight wing on its FJ-1 Fury for the low-speed and handling qualities on landing approach necessary for carrier-based operations.

Swept wings were not new. Most "tailless" or flying-wing aircraft had a swept wing in order to provide enough distance between the pitch-control surfaces and the airplane's center-of-gravity for adequate pitch-control power. One of the Navy's first airplanes was the Burgess-Dunne AH-7, a swept-wing biplane with no horizontal tail. The benefit was the potential for reduced drag and weight by eliminating the aft fuselage and horizontal tail. There were drawbacks, however: pitch sensitivity, a smaller center-of-gravity range, and poorly damped dynamic longitudinal stability. Although other tailless/flying wing designs progressed to flight test, there was virtually no series production of any up through the end of World War II except for a few gliders and the German Me 163, a rocket-powered interceptor.

In the U.S., Jack Northrop was the leading proponent for the flying wing. Although he designed and flew demonstrators and received contracts for both a fighter and a bomber that reached flight testing in WWII, none of his designs achieved operational status.

Many at the Bureau of Aeronautics believed that jets could not be made compatible with aircraft carrier operations. In April 1947, Metsger's boss was still justifying a turboprop replacement of the AD Skyraider in part as insurance against the unsuitability of turbo-jet fighters for combat air patrol, which required endurance. Sweeping the wing with its associated low-speed handling quality problems seemed even less feasible.

In spite of the doubts of his seniors and those in BuAer's Power Plant Division, CDR Metsger was able to pursue an aggressive but well thought-out plan to not only bring the carrier Navy into the jet age, but with airplanes possessing state-of-the-art performance. He alleviated some doubts about low-speed handling qualities by initiating a low-cost, swept-wing demonstration and evaluation program in 1946. The result was the Bell L-39, a P-63 King Cobra modified with swept wings. It provided the confidence that leading-edge slats greatly mitigated the handling quality issues at approach speeds and also provided a reasonable stall speed. The results were a bit misleading in that the poor thrust response of the jet engine was not represented.

The bright promise of the F7U-1 went unrealized for several reasons. Among them were a significant

Top-to-bottom: Jack Northrop was a leading advocate in the United States for tailless aircraft. He received a contract from the Army in June 1940 for a fighter incorporating his design concept. Engine availability and installation problems delayed the flight test program. Once in flight test, unacceptable handling quality issues were revealed, further delaying the program to the point where its inferiority to jet fighter performance resulted in cancellation. (USAF) Northrop and the Army persisted in the development of tailless aircraft because small-scale demonstrators like the N-9M shown here provided the confidence that the concept was valid. Note that these were true flying wings, with no vertical or horizontal tail surfaces. Yaw control was accomplished with wingtip-mounted surfaces similar to dive brakes. (USAF) Closely following on the heels of the XP-56 contract, Northrop received a contract for a bomber, B-35, from the Army in November 1941. First flight of the first XB-35 was not accomplished until June 1946, however, and even if problems weren't being experienced with the engine installation and contra-rotating propeller system, it was clear that the future bombers would be powered by jet engines. (USAF) To evaluate the benefit of jet engines on a flying-wing bomber, three B-35s were converted to substitute eight General Electric J35 engines for the four big Pratt and Whitney R-4360 engines. However, for various and controversial reasons, the program was cancelled. (USAF)

Above, as late as 1947, there were those in the Bureau of Aeronautics who preferred a turboprop-powered airplane for some fighter missions that required endurance as opposed to speed. As it happened, reliance on the Douglas A2D Skyshark for this requirement would have been unfortunate as its Allison turboprop engine defied extensive effort to make it durable and reliable. (Terry Panopalis collection) Below, to alleviate fears about the all-important low-speed handling qualities of the swept-wing for carrier operations, the Fighter Class Desk officer initiated a flight-evaluation program with Bell Aircraft that swept the outer wing panels of a P-63. The L-39 demonstrated that leading edge slats provided acceptable stall and handling-quality characteristics. (NA 80-G-702663)

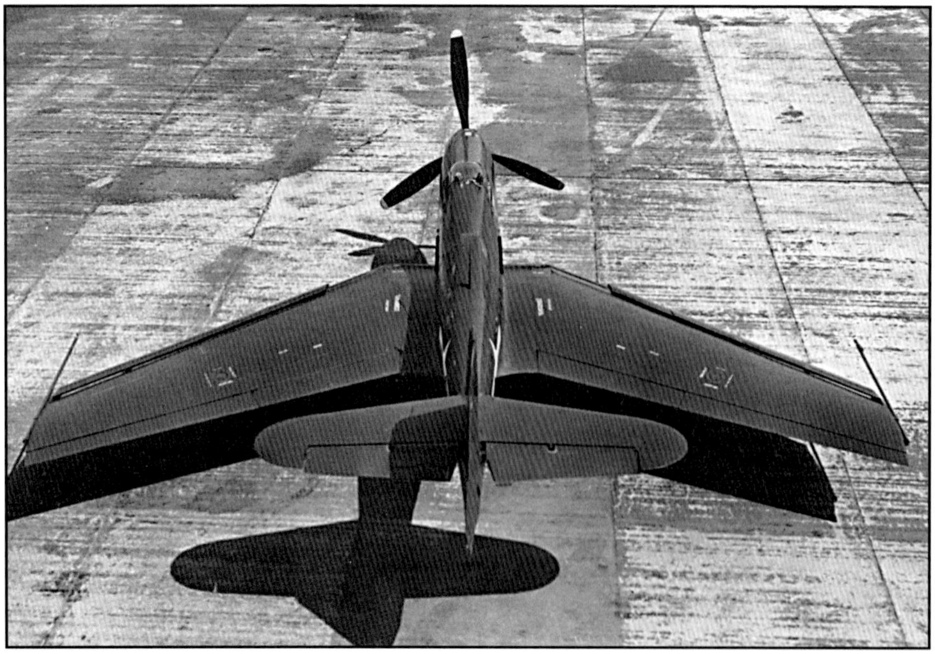

underestimate of the empty weight; Westinghouse's inability to develop, qualify, and deliver the J34 engine on schedule; delays with the new Solar afterburner; handling quality problems in flight test; accidents; a Defense Department-directed budget reduction in 1949; and "better" being the enemy of "good enough".

BuAer's 1946 Day Fighter Competition

The Navy's Bureau of Aeronautics was responsible for procuring and supporting aircraft to fulfill the requirements of the Chief of Naval Operations (CNO). This included contracting for improvements to existing aircraft, engines, and avionics as well as development of new ones. (Weapons development was the responsibility of the Bureau of Ordnance.)

By late 1945, the Navy had already taken baby steps in the development of carrier-based jet fighters. The Ryan FR-1 was in production. It combined a piston engine in the nose with a jet engine in the aft fuselage. The McDonnell XFD-1 was in development, having flown for the first time on 26 January 1945 powered by two Westinghouse 19B engines. Although it would be the first U.S. Navy jet fighter to fly onto and off of an aircraft carrier at sea, that milestone didn't occur until July 1946. BuAer had also evaluated two Army Air Forces projects, the first U.S. jet, the Bell P-59 and the second, the Lockheed P-80, at Patuxent River. The Navy's second jet, the North American FJ-1 Fury, would not fly until September 1946 and the third, Vought's F6U, on 2 October 1946. The improved model of the McDonnell FD, the F2D, would not fly until January 1947. All of these were straight-wing designs.

CDR Metsger wanted to move from baby steps to giant strides on his watch at BuAer. At the end of 1945, he had coordinated the creation of confidential Bureau of Aeronautics Outline Specification (OS) 105 for a day fighter with a maximum speed in level flight of 650 mph (564 knots and Mach .85 at sea level) and an initial rate of climb of 10,000 feet per minute. Service ceiling was to be at least 40,000 feet, but "no premium would be earned for predicted service ceilings of more than 50,000 feet."

The fighter had to have sufficient fuel to accomplish three different mission scenarios:

Fighter Radius: Not less than 300 nautical miles. Engine warm-up for 30 seconds and takeoff for one-minute at rated takeoff thrust; climb to 5,000 feet at military thrust and rendezvous for 10 minutes at speed for maximum range; climb to 30,000 feet at military thrust and cruise at maximum range speed; five minutes of combat power (afterburning or "other means of pro-

ducing additional main engine thrust") and five minutes of military thrust at 20,000 feet; climb to 30,000 feet at military thrust and cruise back at maximum range speed; 15 minutes of loiter at 10,000 feet and 10 minutes at sea level with landing gear and flaps down; and one 30-second military-rated thrust wave-off.

Interceptor Radius: Not less than 100 nautical miles. Engine warm-up for 30 seconds and takeoff for one-minute at rated takeoff thrust; cruise out at V max at sea level at military thrust; five minutes at combat power; cruise back at sea-level maximum-range speed; 30 minutes loiter at sea level; 15 minutes rendezvous at sea level; landing phase as specified for the fighter mission.

Cruising Endurance: At least three hours at speed for maximum endurance at 30,000 feet. Engine warm-up for 30 seconds, takeoff for one-minute at rated takeoff thrust, and climb to 30,000 feet at military thrust; loiter, and landing phases as specified for the fighter mission.

Takeoff stall speed was not to exceed 105 mph, which assumed a catapult end speed of 86 mph, wind over deck of 25 knots (29 mph), and a margin of 10 mph. (Airplanes with a takeoff weight greater than 22,000 lbs were required to have a lower stalling speed because the catapult end speed would be lower.) Stall speed with 25% of internal fuel was to be no more than 90 mph.

When folded, kneeled, etc., the size and shape of the airplane had to be such that at least 25 could be "spotted" or parked on a carrier deck 200 feet long by 96 feet wide. Moreover, a folded airplane was to be accommodated on the carrier elevator without "precise positioning within more than one critical dimension" and be not more than 17 feet in height.

An indication of the importance that the Navy placed on deck-space utilization was the statement "Should the contractor find it possible to exceed the performance requirements, emphasis should be directed first toward reducing the size of the aircraft and second, toward improving the maximum speed."

A pressurized cabin was required that maintained a pressure equivalent to 5,000 feet up to an altitude of 17,500 feet and a pressure differential of 2.75 psi above that. The contractor was to provide a "reasonably safe emergency egress for the pilot under conditions of high speed and possibly uncontrolled flight." To ensure that there was no misunderstanding about the means desired, the specification made it clear that "positively ejecting the pilot from the airplane" was considered necessary.

Although armament was to be four 20mm guns with 250 rounds of ammunition for each gun, a study was desired that indicated "the

Above, the Navy initiated the Ryan FR-1 program in February 1943 as a low-risk way to combine jet-like performance with the short takeoff and landing capability necessary for carrier operations. The Fireball was powered by a Wright R-1820 in the nose and a General Electric I-16 (J31) in the tail. FR-1 pilots were often tempted to pull up alongside other airplanes, feather the propeller, and maintain position using the jet. Since relatively few had seen a jet airplane at the time, this was an attention-getter. (NMNA) Below, only a few months after the McDonnell XFD-1 Phantom flew for the first time, the Navy ordered the very similar F2D-1 Banshee that was powered by higher thrust engines for better performance. The relatively trouble-free development of the FD proved misleading as encounters with compressibility required extensive modification to the F2D's wings and empennage. (NA 80-G-395929)

Above, the Army provided a few Bell P-59s, the first U.S. jet airplane, to the Navy for its familiarization with the performance and operation of jet fighters. Its marginal speed advantage over propeller-driven airplanes did not deter either the Army or Navy from funding jet airplane programs at other companies. (National Museum of Naval Aviation) Below, the Navy came close to buying a carrier-compatible version of the Army's first operational jet, the Lockheed P-80. One was modified with catapult and arresting hooks and flown to and from Franklin D. Roosevelt in November 1946. The Navy, however, needed much better low-speed capability and elected to continue with its own programs. (Hal Andrews collection) Bottom, Vought received a letter of intent for its first jet airplane from the Navy in December 1944. Powered by a single Westinghouse J34, the XF6U first flew in October 1946. Although the Navy did eventually buy 30 production Pirates, it was too little, too late from an operational standpoint and was never carrier-qualified. (NA 80-G-703822)

changes necessary to accommodate a change to four 0.60 cal. fixed guns with 250 rounds of ammunition per gun."

Because engines took longer than airplanes to develop—and once developed usually powered several different airplanes—the military services contracted directly with the engine manufacturers for development and production programs and provided engines to the airplane manufacturers as government-furnished equipment.

ANA Bulletin 306 dated January 1947 assigned official U.S. government designations to jet engines for the first time. Up until then, a jet engine was identified by its manufacturer's designation. The engines were numbered roughly in order of their development, with Navy engines given even number designations and Army engines, odd numbers. For some reason, engine numbering began with 30. The complete designation included a one or two-letter identifier of the manufacturer of the engine, J34-WE indicating that Westinghouse built the engine, for example.

In 1946, the following were available for consideration (ANA Bulletin 306 designations provided in parentheses):

Engine	Thrust (lbs)	Air Flow
Westinghouse 19B (J30)	1,680	Axial
Westinghouse 24C (J34)	3,000	Axial
General Electric I40 (J33)	4,000	Centrifugal
G.E. TG-180 (J35)	4,000	Axial
Lockheed L-1000 (J37)	5,000	Axial
Rolls-Royce Nene (J42)	5,000	Centrifugal

While stating that "The contractor's choice of engines is unlimited within those gas turbines, turbo-jets, rockets, or ram-jets under development or manufacture in the United States," BuAer recommended proposals powered by the Westinghouse 24C and the General Electric I40 and TG-180 engines due to schedule considerations. (The fact that the General Electric engines were developed on Army contracts didn't matter.) Although Westinghouse had not yet qualified the 24C (J34) engine, it was considered to be low risk because it was a scale-up of their successful 19B (J30) and had first run in April 1945. The 24C was also slated to power the relatively small single-engine Vought F6U, twin-engine McDonnell F2D, and some Air Force jet fighters as well. North American had been directed to use the TG-180 (J35) in the FJ-1 Fury and Republic had selected it for the P-84. The I40 (J33) was in production for the Lockheed P-80. Neither the Nene (J42) or the L-1000 (J37) were likely to be selected since the former was not yet in production in the US and the latter had not yet run.

In addition to the development status, another engine selection consideration was that, absent an auxiliary-thrust feature like a rocket or ramjet, at least two engines of the size then available would be needed to provide the thrust for the high-speed and rate-of-climb performance requirements. The centrifugal flow I40 was hard to accommodate in a twin-engine configuration because of its diameter, effectively ruling it out. Although two 24Cs only produced 6,000 lbs of thrust compared to 8,000 from two TG-180s, the 24C had a better thrust-to-weight ratio and specific fuel consumption in cruise than any of the other engines. Afterburning, which was predicted to increase sea level static thrust by up to 30%, more than made up the difference in thrust if 8,000 lbs was adequate for combat power.

Afterburner Development

The jet engine compressor stuffs high-pressure air into a combustion chamber, where it is mixed with fuel that is set alight. The expanding hot gas escapes out the tailpipe, providing thrust. The exhaust also turns a turbine that turns the compressor, completing the cycle. This turbine has to withstand very high temperatures. Development of material that was durable enough for the purpose was one of the reasons that jet engines did not become available sooner. Even after metallurgy provided an initial solution with very low times between overhaul, turbine blade thermal limits meant that only about 25%

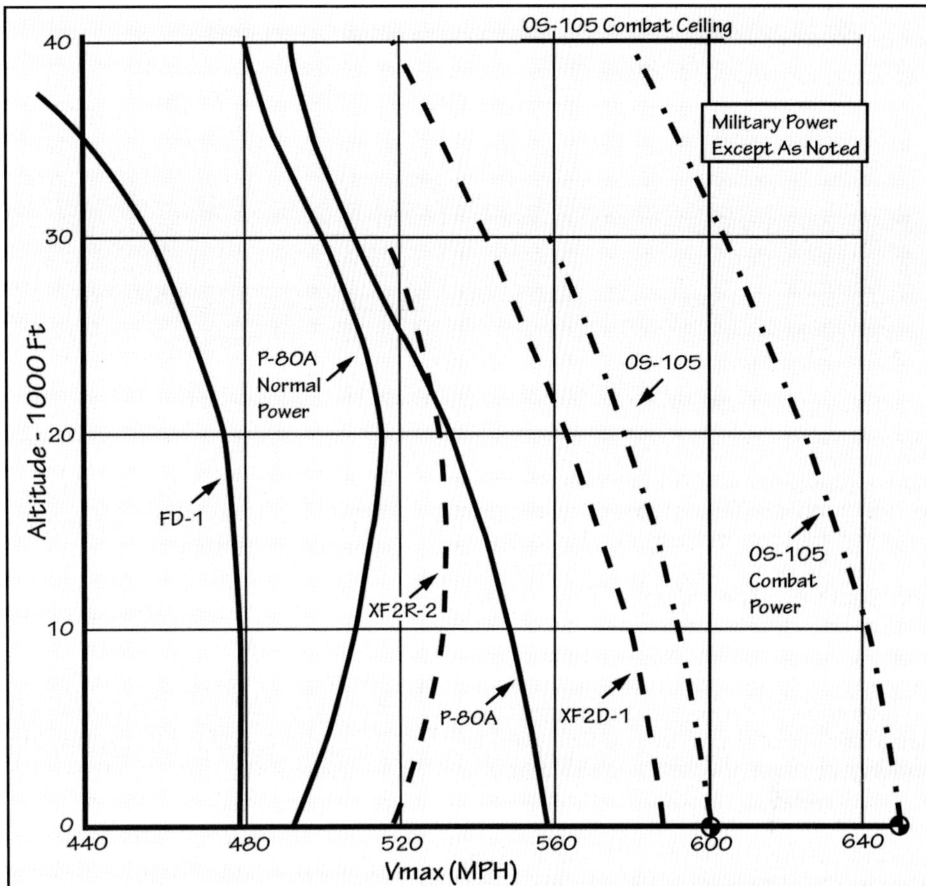

Early 1946 Performance versus OS-105 Requirement

BuAer's Fighter Class Desk was clearly trying for a great leap forward in early 1946. The Navy's first operational jet, the McDonnell FD-1, was inferior to the Air Force's Lockheed P-80A. Its successor, the F2D-1, was to be superior but had not yet flown. OS-105 required a top speed of 650 mph, about 10 percent more than the F2D-1's, through the use of unproven afterburners. The XF2R-2 was an attempt to achieve pure-jet performance with the turboprop-powered F2R-1 by substituting a more powerful jet engine and lower drag jet intakes; it was canceled. (Author) Below, one of the first concepts for a variable-area nozzle was a partial plug mounted on the end of a lance that could be moved back and forth within the tailpipe. When extended aft, it reduced the exit area so it was suitable for basic engine operation. When pulled forward, it opened up the exit area for the additional thrust of the afterburner. This was the configuration Vought used in its V-346 proposal as indicated in drawings by the cones extending from the tailpipe. (Author)

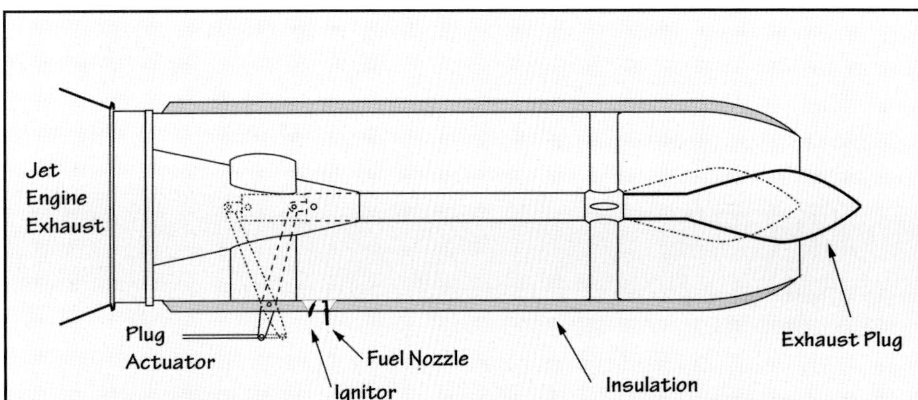

Early Variable Nozzle Afterburner Concept

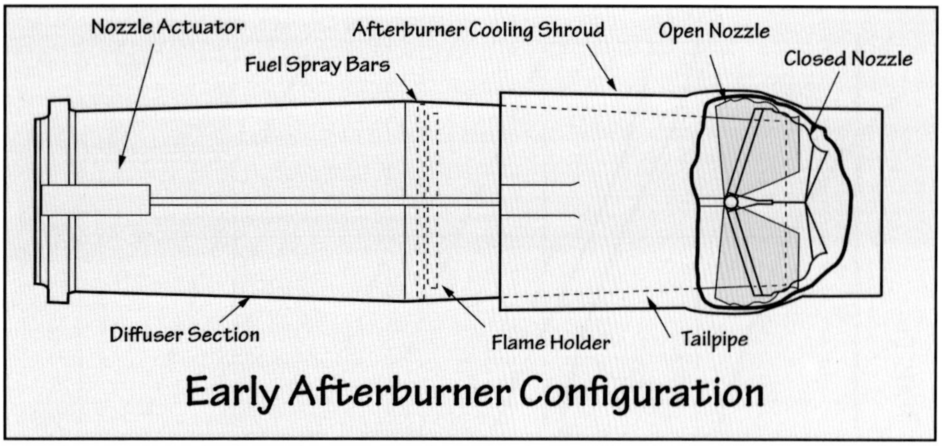

Above, the Solar afterburner was designed to bolt onto the tailpipe of an existing jet engine. The long diffuser section reduced the velocity of the exhaust in the combustion section. The variable area exhaust nozzle kept the backpressure on the turbine to a reasonable level when the afterburner was operating. (Note that the actuator was located far forward of the hot section.) The shroud provided a flow path for cooling air over the hot section and protected the aircraft structure from excessive temperature. (Author) Below, this is a view looking forward from the aft end of Solar Aircraft Co. afterburner tailpipe for a J34-WE-30 engine. The "eyelid" variable-area nozzle is fully open. The assembly of troughs is the flame holder. The fuel spray bars are a few inches behind it. This hardware somewhat reduced the thrust of the basic engine. (Author)

of the oxygen in the air flowing through the combustion chambers could be burned.

One of the drawbacks of the jet engine was that its thrust at low speeds was very inferior to that provided by a propeller. It didn't take long for propulsion engineers to realize that the remaining oxygen in the exhaust could be burned to generate additional thrust for takeoff, climb, and combat. For the short periods of time that maximum thrust was required, a smaller engine equipped with an afterburner would therefore be capable of developing the thrust of a bigger (and therefore heavier, more expensive, and less fuel efficient in cruise) engine that did not have one.

NACA Lewis Flight Propulsion Laboratory did initial afterburner experiments in 1943. Abe Silverstein of Lewis designed one based on ramjet engine technology in 1945. It was successfully demonstrated in March 1946 using the Lewis Altitude Wind Tunnel that allowed operation at simulated flight conditions. In England, Frank Whittle's company, Power Jets, was developing a "reheat jetpipe" for an engine to power a proposed supersonic research aircraft. The Germans had also ground tested a Jumo jet engine with an afterburner before the end of the World War II. Nevertheless, in 1946 the afterburner was still not much more than a theory.

On the other hand, an afterburner was even simpler mechanically than the jet engine. All that was required was a fuel spray device, ignitor, and flame holder in the tailpipe downstream of the turbine; a variable-area exit nozzle (the only moving part) to reduce the back pressure when the afterburner was running; and an externally cooled tailpipe made of temperature-resistant material. The problem, as one engineer described it, was that afterburner ignition was "like lighting a bonfire in a hurricane." The additional thrust was also accompanied by a prodigious increase in fuel consumption.

The afterburner had two notable drawbacks in addition to prodigious fuel consumption. First, engine thrust when the afterburner was not in use was slightly reduced by the presence of the framework that housed the afterburner fuel injection manifold, igniter, and flame holders: for the J34, the loss in thrust was originally estimated to be about 160 pounds per engine, or 5%. Second, it increased the airplane's empty weight because of the weight of the components, not least being the heat-resistant and air-cooled tailpipe that still limited continuous afterburner operation to five minutes.

CDR Metsger wrote a memo in March 1946 justifying the development of a 7,000-lb thrust jet engine in which he expressed his concerns about the afterburner. He cited a Douglas study associated with its night-fighter proposal that resulted in the F3D Skyknight. It concluded that for the VF(N) mission, a jet airplane utilizing afterburning was inferior to one that did not. He also noted "afterburning has not yet been worked out in a flight article." That was an understatement. In fact, the amount of thrust that would be added by the afterburner, not to mention the fuel consumption, was uncertain. Vought engineering estimates of the thrust increase varied significantly depending on the fuel/air combustion ratio that could be achieved: about 20 to 25% at sea level and zero airspeed and 42 to 48% at 20,000 feet and 500 mph. (An estimate of 23% at sea level was used for preparation of the XF7U-1 SAC (Standard Aircraft Characteristics) chart dated 1 February 1947.)

Above, the prototype Solar afterburner was as long or longer than the J34 engine. Here one is being fit-checked to a J34 at Vought prior to installation in XF6U BuNo 33533, which has been modified with an extended aft fuselage to accommodate it. The variable area nozzle is located inside the cooling shroud, making the tailpipe look a little like an undigested meal in a python. (Vought Archives) Below, the Vought F6U was originally designed without an afterburner. One was subsequently incorporated for production in order to restore performance lost due to weight increase in development and redesign to add an ejection seat and pressurization. While it did not make the F6U a success, Vought and the Navy benefitted from the design and development experience. (Author from Vought Archive photos)

Still, the increase in thrust provided by "after burning", as it was initially called, was seductive. It provided notably higher acceleration, rate of climb, and top speed. The projected thrust increase percentage also increased with airspeed and to a lesser extent, with altitude. Because afterburning had the potential to provide a significant increase in maximum thrust at a relatively small empty-weight penalty, both Ryan Aeronautical Corporation and Solar Aircraft Company received contracts from BuAer in 1946 to design and develop afterburners. While the afterburner proved to provide the desired thrust improvement, development and qualification was more difficult than hoped, affecting a generation of Navy fighter programs. Some still believed it to be counterproductive for a carrier-based day fighter as late as the early 1950s, including the officer who headed the BuAer fighter class desk at that time.

The Navy had issued the development contract for the F6U, Vought's first jet, in December 1944. It was originally designed and built without an afterburner. It first flew on 2 October 1946, powered by a single Westinghouse 24C engine. The F6U had already proved to be short on thrust versus weight before BuAer

Before After

Vought XF6U-1 BuNo 33532 Afterburner Modification

Above, Vought's initial V-346 study, accomplished before the OS-105 requirement was released, was smaller and lighter than the eventual proposal V-346A. Its Westinghouse J34 engines were non-afterburning. Given the program delays that resulted in part from afterburner development, BuAer might have deployed a swept-wing day fighter years earlier if it had not tried for such a great leap forward in performance with OS-105. (Author) Below, the V-146 resembled some WWII German flying wing designs that may have gotten as far as wind tunnel test. The Junkers EF 128 shown here was one. However, the war ended before any got so far as a prototype. Vought engineering insisted that the V-146 was not based on a German design. (Mirko Lucini)

decided that jet fighters needed cockpit air-conditioning/pressurization and an ejection seat, among other new design and operating requirements that the XF6U was not designed to. BuAer and Vought hoped that the addition of an afterburner would restore the intended performance. In early 1948, the first flight-worthy Solar afterburner was delivered to Vought for installation in XF6U BuNo 33532. Flight test began in April 1948.

The performance increase was as dramatic as hoped for. At 12,000 feet, the afterburner resulted in an indicated airspeed change from 345 mph to 440 mph with no change in engine operating rpm. As expected, the increase in fuel consumption was even more dramatic. At military power, fuel consumption was 3,744 lbs per hour; in afterburner it more than tripled, to 12,850 lbs per hour, which would empty the F6U's tanks in only 20 minutes. Moreover, afterburner lights and sustained operation were problematic, particularly on takeoff and at altitude. The afterburner therefore did not save the F6U program. Only 35 production Pirates were built. It was destined for a very short career and never operated from a carrier's deck. While disappointing, it provided Vought and the Navy with their first experience with the afterburner from a design and flight evaluation standpoint.

OS-105 Proposals

BuAer provided the Model VF Airplane Request for Proposal (RFP) dated 25 January 1946 to 12 aircraft manufacturers that it considered qualified to do the program. Technical submittals were due on 15 April and cost proposals, on 1 May.

Paul Baker was Vought's chief engineer; F.N. Dickerman, assistant chief engineer; and William Schoolfield, chief aerodynamicist. They had already been working on a small twin-engine, tailless fighter design, the V-346, which was similar in configuration to Vought's ultimately successful proposal. (It resembled some WWII German projects, but they insisted that it was in no way influenced by German research.)

Unlike most of the so-called tailless designs with vertical fins, the V-346 had them located about halfway out on the trailing edge of the wing instead of at the wingtips or on the fuselage.

The pre-RFP V-346 was to have a takeoff gross weight of 15,180 lbs with 820 gallons of fuel and a combat gross weight of 13,200 lbs. The wingspan was 34 ft and wing area, 450 sq ft. It was only 32 feet long because at this point, its J34 engines did not have "after burning."

One V-346 feature that didn't survive the predesign phase was a kneeling capability. The Navy was briefly enamored with the concept, which was to retract the nose gear for parking, raising the airplane's tail high in the air. A small wheel would extend from or be inserted in the forward fuselage in its place to support the nose for taxi and parking. There were three benefits: the nose of one aircraft could be parked under the aft fuse-

lage of another, reducing the amount of deck space required while possibly obviating the need for wing folding; the hot jet exhaust would be directed upward and not at airplanes and deck crew close behind the jet; and the climb up to the cockpit would be short. The FJ-1 Fury and the F6U-1 Pirate were designed with this feature as was the F2H-1 Banshee, although it also had folding wings. Initial operational evaluation of the FH-1 Phantom—which did not have kneeling capability—established that jet blast, at least during taxi, could be accommodated. Not as many airplanes could be parked with kneeling compared to wing folding and the benefit was not significant when combined with wing folding.

The V-346 major change from the pre-RFP design study was the incorporation of afterburners and the addition of more fuel. The fuselage was lengthened accordingly. Like the XF6U, the original pre-RFP design study doesn't appear to have included an ejection seat and probably didn't have pressurization, both now requirements for Navy jet fighters. The wing area grew to maintain an acceptable wing loading at the higher weight.

The inexorable growth in size and weight had begun. Unfortunately, the increased thrust that Westinghouse committed to for each bigger and more capable Cutlass was seldom fully realized, at least with acceptable durability.

Kneeling was an innovation that was intended to increase the number of jets that could be accommodated on the carrier deck or in its hangar. It also directed the exhaust blast up over personnel and airplanes.Above, on 21 May 1947, XF2D-1, BuNo 99858, with original vertical and horizontal tail knees behind XF2D-1, BuNo 99859, with modified horizontal tail to demonstrate the increased deck loading available through this concept. (via Fred Roos) Below, the 3rd XFJ-1, BuNo 39055, fitted with the small kneeling nose gear wheel is seen at the beginning of kneeling process. On the XFJ-1 the nose gear was slowly retracted and the nose section would settle on the kneeling wheel. (North American) Bottom, In the initial Vought design study, kneeling was to be accomplished by partially retracting the nose landing gear. (Author)

Kneeling Position for Taxi and Parking

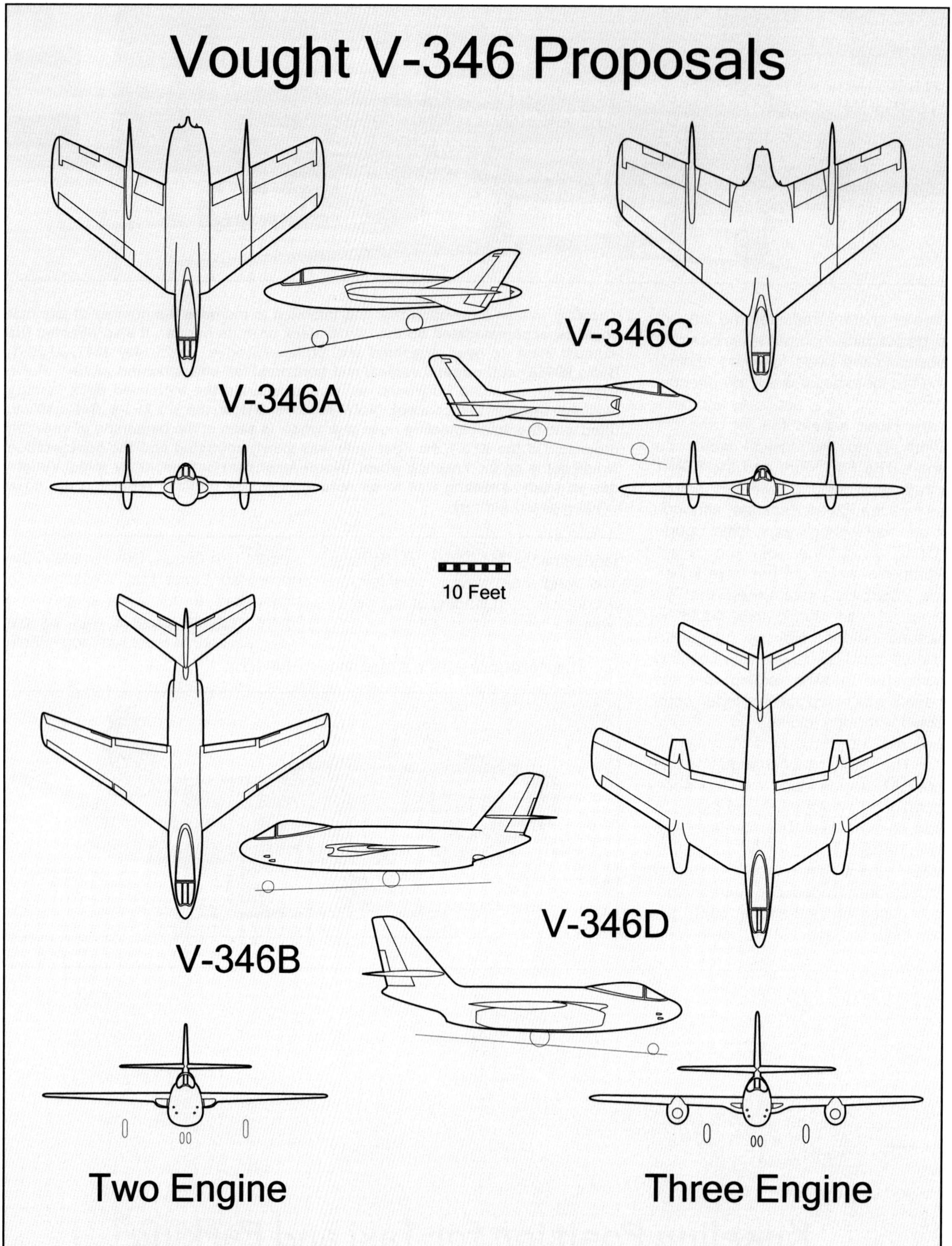
Vought V-346 Proposals
V-346A
V-346C
10 Feet
V-346B
V-346D
Two Engine
Three Engine

At left, in order to be sure that one of its submittals would find favor at BuAer, Vought proposed four different configurations against OS-105, two twin-engine and two three-engine, with one of each being a flying wing and the other a conventional platform. The performance comparisons also reinforced the superiority of the tailless configuration. (Author) At right, Vought's February 1946 V-346 design study was aerodynamically very similar to the V-346A that was proposed in March. Unlike a true flying wing as exemplified by Northrop's designs, it featured large vertical fins and rudders as well as a fuselage. In this study, the tailhook was attached at the extreme aft end of the fuselage and stowed upside down in a recess on the top of the fuselage. (Author)

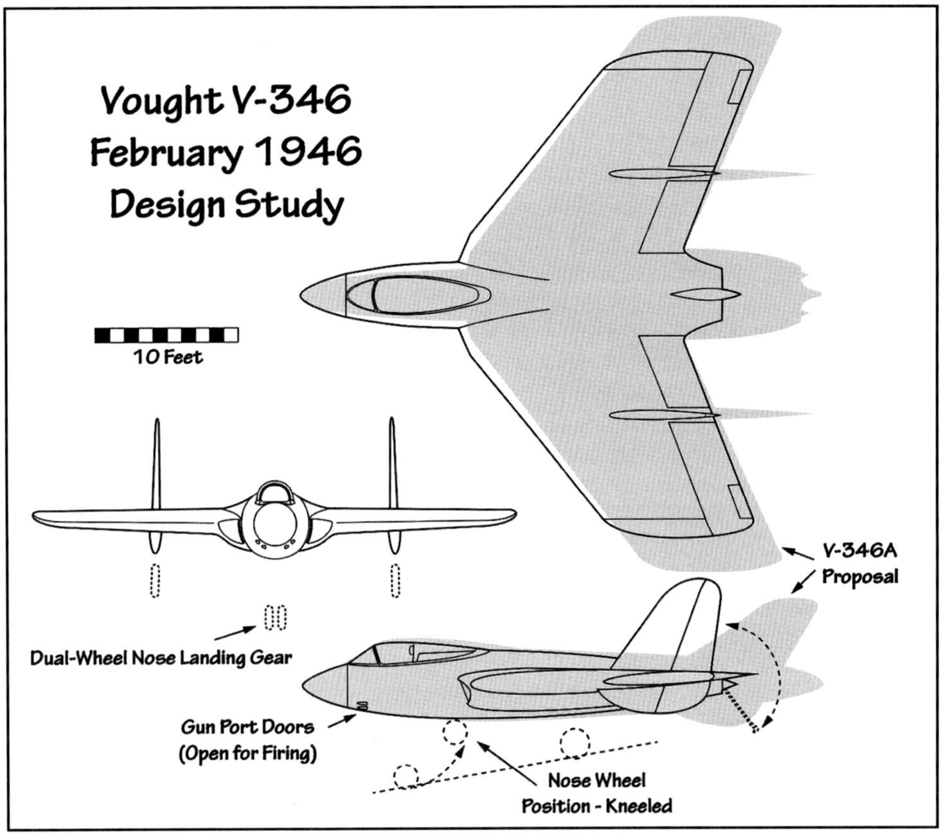

Vought believed that its twin-engine tailless design was superior to the other alternatives that it studied, but in case the Navy did not agree, it submitted four proposals in all, each powered by Westinghouse 24C (J34) engines with afterburners. Two were tailless swept-wing designs and two were conventional swept-wing planforms. Each was identified with the 346 design number along with a unique letter. Two engines powered the tailless design V-346A and the conventional V-346B. Three engines powered the tailless V-346C and conventional V-346D. The V-346C had all three engines mounted in the fuselage, fed by two inlets, so it was very similar in layout to the V-346A. The V-346D had only one fuselage-mounted engine; the other two were wing-mounted and widely spaced.

McDonnell, Curtiss-Wright, North American, and possibly Boeing submitted two designs each. Douglas and Martin each proposed only one. Grumman apparently did not propose, possibly because it was in the process of negotiating a contract with BuAer for the XF9F-1, a four-engine night fighter. The proposed engines were either the Westinghouse 24C (J34) or the General Electric TG-180 (J35).

Of the four Vought designs, the tailless V-346A was the lightest and therefore had higher performance than the other twin-engine design, the swept-wing V-346B. As would be expected, the three-engine V-346C was heavier than the two-engine designs but had better performance because of the greater installed thrust to weight. The three-engine V-346D was the heaviest and slowest of the four, although its projected performance still exceeded the OS-105 requirements. The only shortfall among the candidates was the V-346B's maximum rate of climb at military thrust.

In those simpler times, BuAer accomplished its evaluation of the technical proposals and had selected one only six weeks after they were submitted.

In BuAer's request for authority to contract dated 13 June 1946, it provided the following summary of the fixed-price proposals that were received from six manufacturers*:

	OS-105 Requirement	V-346A	V-346B	V-346C	V-346D
Combat Gross Weight (lbs)		14,300	15,700	17,500	20,950
Max Speed (mph)*	650	666	650	670	652
Max Speed (mph)**	600	620	605	655	610
Max Rate of Climb (fpm)*	10,000	11,430	11,250	12,320	12,310
Max Rate of Climb (fpm)**	6,500	6,520	5,800	9,670	7,000
T.O. Rate of Climb (fpm)	500	830	630	1,390	1,110
W.O. Rate of Climb (fpm)	500	1,730	1,470	2,510	2,020
T.O. Stall Speed (mph)	105	105	105	105	102
Landing Speed (mph)	90	89.3	89.2	89.6	87.1
Service Ceiling (ft)**	40,000	44,400	43,300	50,000	46,400
Fighter Radius (nm)	300	300	314	318	300
Interceptor Radius (nm)	100	100	105	102	100
Cruising Radius (hours)	3	3	3	3	3

***Combat Thrust (with afterburning) and combat weight**
**** Military Thrust (maximum without afterburning) and combat weight**

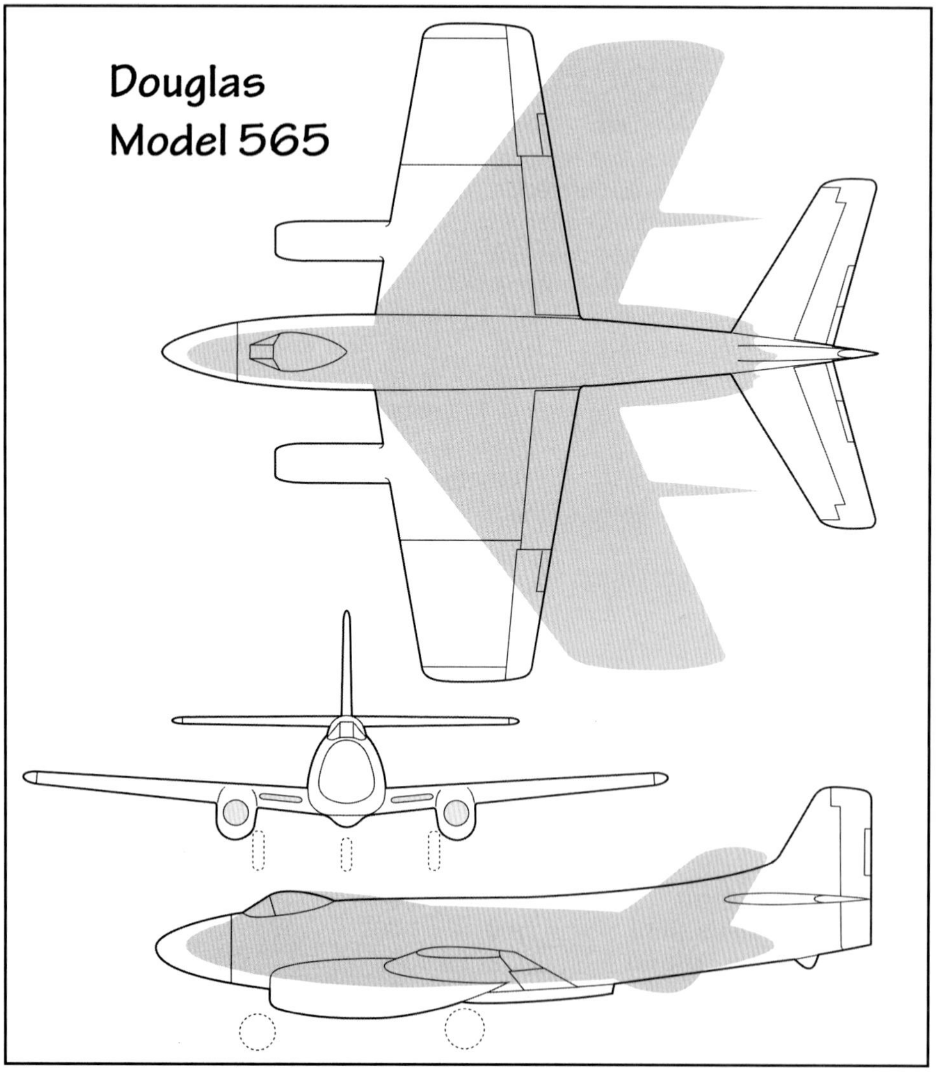

Manufacturer	Design	Total Price
Curtiss-Wright	#515	$11,100,429
Curtiss-Wright	#516	$11,154,416
Chance Vought	#346-A	$8,156,957
Chance Vought	#346-B	$8,040,259
Chance Vought	#346-C	$8,847,157
Chance Vought	#346-D	$9,560,427
Douglas	#565	$9,204,346
Glenn L. Martin	#235	$7,702,517
McDonnell	#40	$7,990,000
McDonnell	#40-A	$7,990,000
North American	2-engine	$9,166,565
North American	3-engine	$11,417,085

*Boeing reportedly submitted proposals for their models 454 and 459. If so, BuAer did not include them in its summary, perhaps because they were not in complete compliance with the invitation to propose.

The Vought Sales Manager, in an internal memo documenting a discussion with one of his BuAer contacts, reported that Vought's twin-engine tailless V-346A was ranked first and the V-346B, second. Note that neither was the lowest price, with the V-346A being slightly more expensive than the V-346B. In order of proposal ranking by contractor, Douglas was second followed by North American, McDonnell, Martin, and Curtiss-Wright. According to him, the Navy thought the proposed Curtiss-Wright airplane to be "an extremely bad design."

Top-to-bottom: Martin's Model 235 was a twin-engine design that was similar in configuration to its B-51 with a podded engine under the fuselage and one in the tail. It was an also-ran in the Navy's evaluation. (Stan Piet collection) The straight-wing D-558-1 was used to set a speed record of 650.7 mph in August 1947, demonstrating that swept wings and afterburning were not essential for high-speed flight. (80-G-703822) In addition to being the only proposal with a straight wing, the Douglas Model 565 was reportedly the only one that did not rely on afterburners for high-speed and climb performance; instead, it was powered by three non-afterburning 24C engines, two in nacelles under the wing and the third in the mid-fuselage. One or more engines was shut down as required to achieve OS-105 range and endurance. (Author)

According to Hal Andrews, the Douglas Model 565 was the only proposal to have a straight, rather than swept wing. In June 1945, the Navy had contracted with Douglas for a three-phase, high-speed research program. Phase I was to result in a jet-powered airplane, the D-558, with a thin straight wing; Phase II was to add rocket propulsion for supersonic flight; Phase III was the preliminary design, including a mockup, of a fighter to be based on the Phase I airplane. In January 1946, Phase II was changed to be a swept-wing airplane following review of German research. (At some point, Phase III was deleted.) However, Douglas' chief engineer, Ed Heinemann, may have felt that a swept-wing design was too risky to propose for OS-105 given the state of the art at that time. (He was, however, in the process of reviewing German aeronautical research reports that would be the genesis of the quasi-delta F4D Skyray.) Andrews wrote that it was a finalist along with the Vought designs because it was considered a good and conservative approach: "most simple and straight-forward design, offering relatively high performance with a minimum of unknowns." However, it didn't meet the top speed or maximum rate-of-climb requirements and the V-346A did.

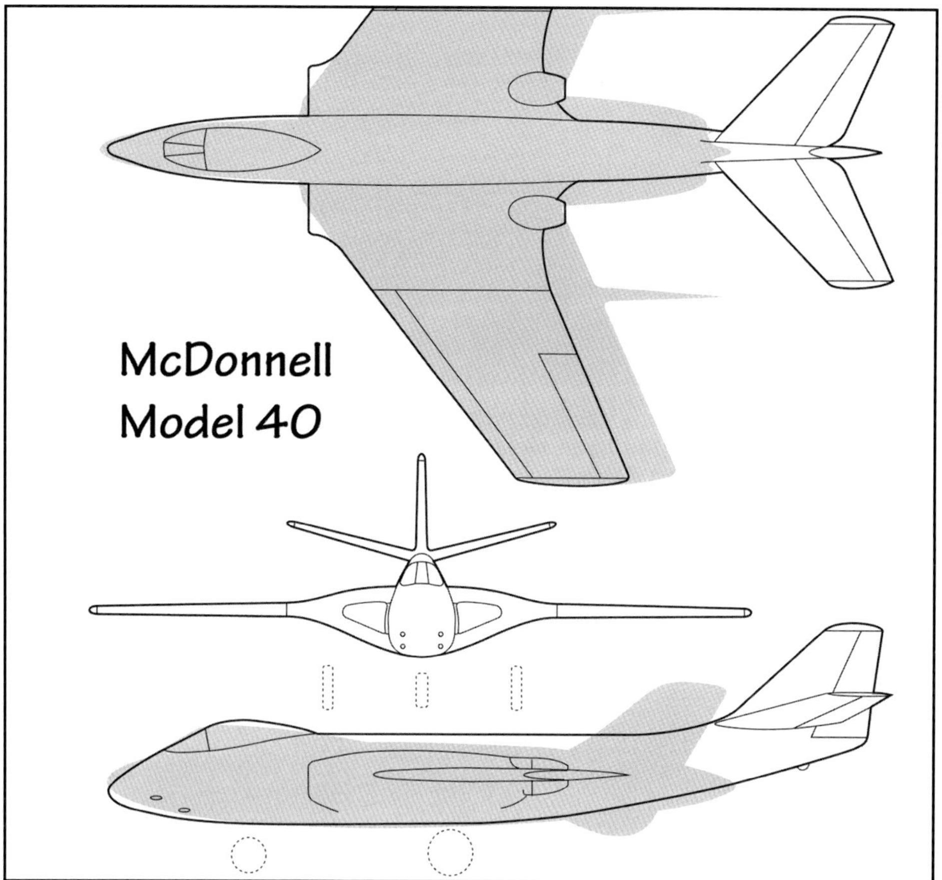

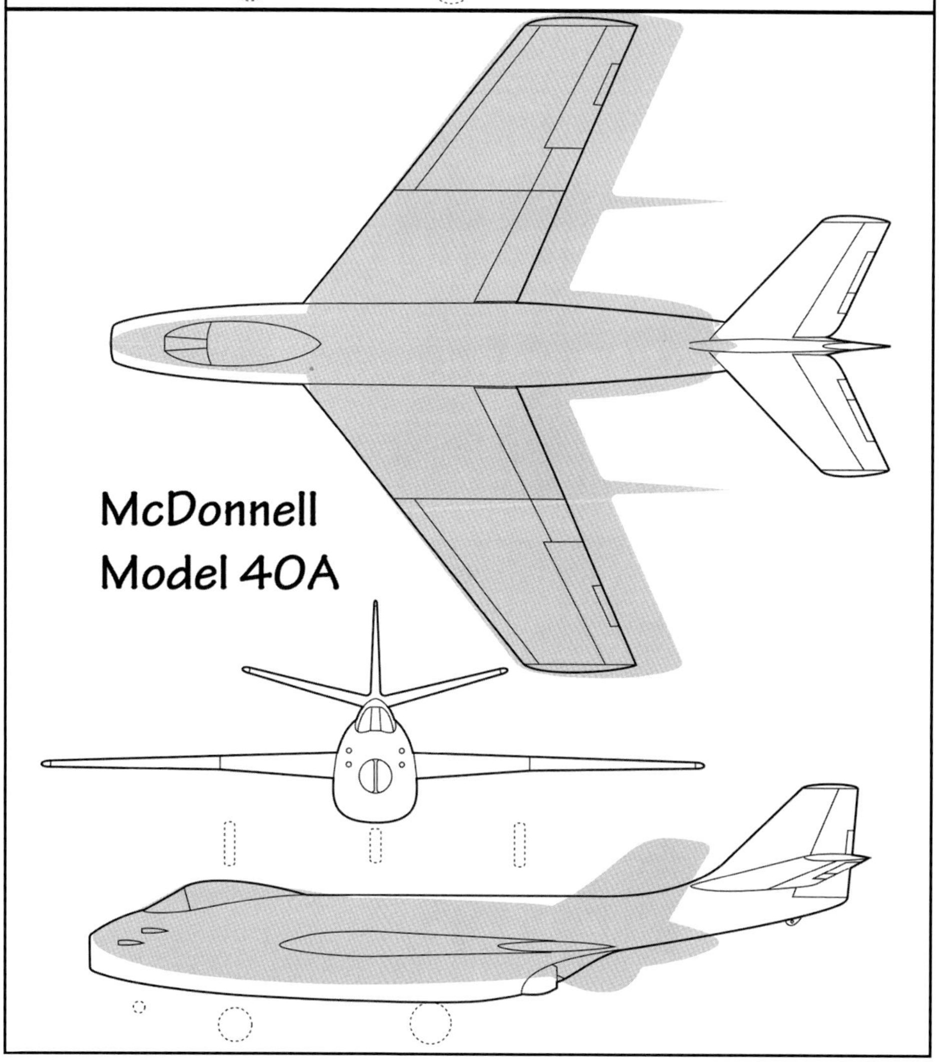

At top right, McDonnell submitted two proposals, both powered by two afterburning J34 engines. The Model 40 was similar in configuration to its FD and F2D designs with the engines located in the wing roots. The cockpit was well forward to provide a view over the nose of 16 ½ degrees. All the fuel was located in the fuselage. (Author) At right, McDonnell's alternate proposal located the engines in the fuselage with the intakes in the nose. As with the Model 40, the afterburners are much shorter than the Solar or Ryan designs (McDonnell did eventually design and flight test a short afterburner on the F2H Banshee.) Kneeling capability was featured, apparently more effective in this configuration because the aft fuselage was upswept. (Author)

Above, Vought submitted a desktop model of the V-346A along with its proposal on 15 April 1946. Note the large clear canopy with excellent rearward visibility. (Vought Heritage Center) Below, Vought decided not to incorporate "kneeling" to provide nose-under-tail overlap because the unique V-346A configuration and stance allowed an overlap of the nose over the wing. This provided the compact parking benefit of kneeling without the additional hardware and deck crew effort. (Author)

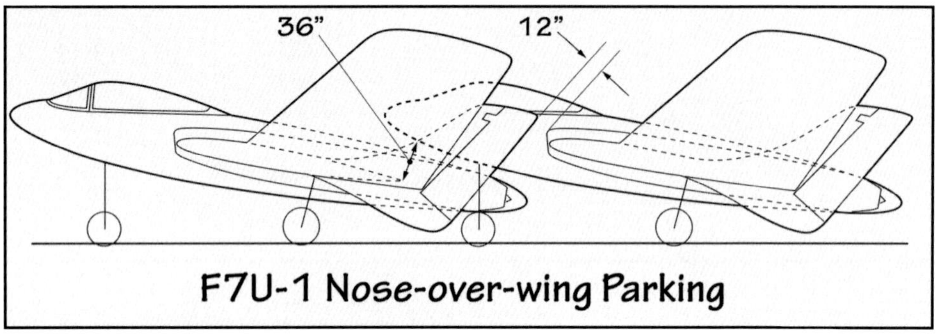

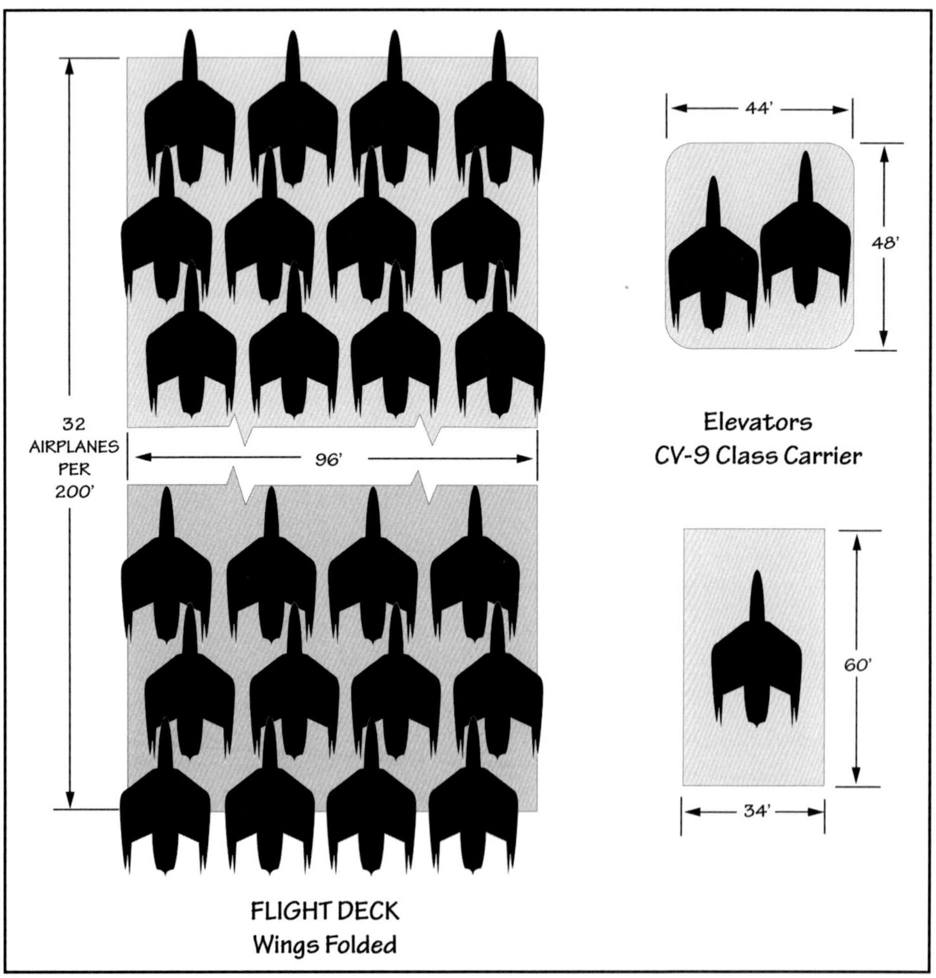

In addition to being lighter and having higher performance, the tailless V-346A was also attractive because of the transonic handling quality concerns. In general, increasing aft stick or trim was required to keep the airplane's nose up due to what was known as Mach tuck. The pilot also had to take care in some airplanes when decelerating through this speed range to avoid a violent pitch-up. This was due in part to the change in air flow aft of the shock wave on the wing causing ineffectiveness of the horizontal tail. In George Spangenberg's oral history, he recalled that Walter Diehl, BuAer's chief aerodynamicist, said at the time "at least we'll solve the subsonic tail buffet problem."

V-346 Design Philosophy

Vought submitted a densely worded, 20+ page Design Philosophy along with its OS-105 proposal. It documented the benefits of the tailless, twin-engine V-346A proposal and its superiority compared to conventional designs for a high-speed, carrier-based jet fighter. It was primarily qualitative (there was only one diagram) so it wouldn't be too challenging for Naval officers with a limited engineering background to understand. For the more technically oriented at BuAer, a bibliography was included that listed "published literature of interest concerning tailless airplanes which have been built in the past, and also a considerable amount of general interest on the subject which has been published by the NACA."

At left, from the standpoint of basing an airplane on an aircraft carrier, smaller was better. There were two standards: how many could be parked in 200 ft of a 96-ft wide carrier deck and compatibility with the Essex-class elevators. Due to the small size of the V-346A, 32 could be parked in the specified deck space versus the minimum requirement of 25, a significant advantage. No "angling" was required to position the airplane on the elevators. (Author)

Vought's design goal was the lightest and smallest airplane that met the mission requirements. Rockets, ramjets, and the "turbo-propeller" were rejected out of hand. "Of the available turbo-jet engines, the Westinghouse 24C best fulfills the desired specific weight and specific volume criteria." Two engines, with afterburning, were determined to meet the performance requirements.

The design philosophy addressed both the benefits and risks of the "tailless" configuration (even though it had vertical tail surfaces, Vought—"for the sake of brevity"—referred to it as such). Based on comparing tailless and conventional designs to the OS-105 mission requirement, it concluded that the tailless design was at least 10 percent lighter than the conventional ones, which translated to better rate of climb and higher top speed. It was also smaller. From an aerodynamic standpoint, the tailless design avoided destabilizing effects of jet thrust, buffet on the tail from open air brakes, and the buffet and trim changes resulting from the influence of the wing on the horizontal tail in high speed dives.

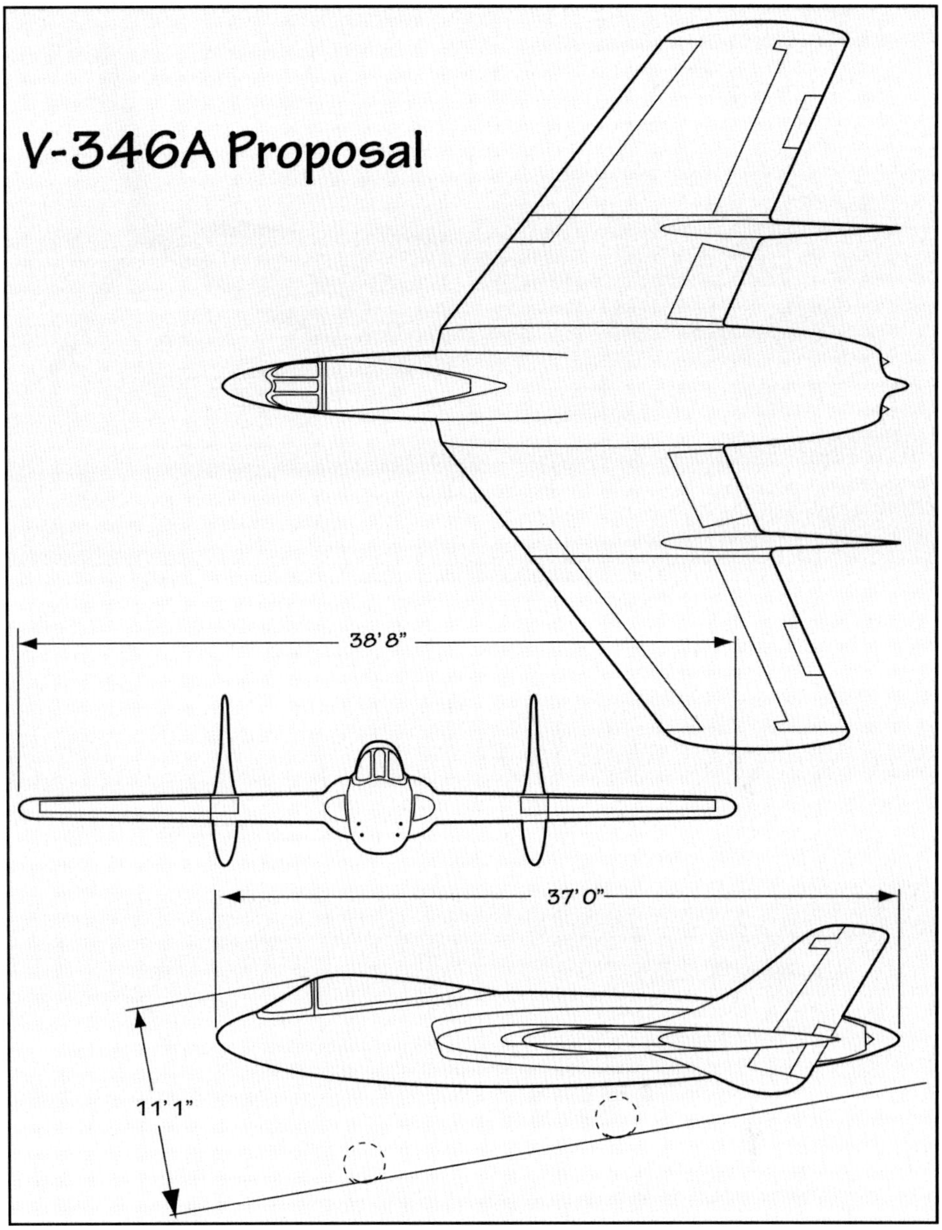

In anticipation of concerns about inadequate pitch damping of a tailless design, Vought cited NACA documents that stated this was not a problem "provided that sufficient static longitudinal stability is present." Similarly, the philosophy addressed the issue of lateral and directional stability using published NACA documents. According to the philosophy, a possible problem with excessive dihedral at high angle of attack would be dealt with, if necessary, by incorporating negative dihedral at the wing tips like some tailless designs. Spin recovery was judged to be a non-issue with an appropriately low aspect ratio and the presence of leading slots on the wing. The prospect of tumbling was dealt with by citing an NACA report on stability and control of tailless airplanes: "There are no authenticated instances of the occurrence of tumbling in flight." While this report stated that models of tailless aircraft could be made to tumble in the wind tunnel "by forcing the model to simulate a whip stall," Vought dismissed this with the statement "It seems extremely unlikely that it will be possible for an airplane to tumble unless it possesses an unstable variation of pitching moment about its center of gravity over a substantial range of angles of attack." A simple experiment with a 3x5 card and paper clips was recommended to the reader if he or she was unconvinced.

In accordance with high-speed flight design theory at the time, Vought selected a low-aspect-ratio (3.0), swept (35° at the quarter-chord) wing for the tailless designs. (The conventional V-346s had a wing sweep of only 30° for lower weight but the tailless designs needed more for control effectiveness.) Vought proposed a symmetrical 10.4% (12% when measured perpendicular to the quarter chord) thick airfoil, with the maximum thickness at 40% wing chord and a trailing edge angle of 12°. German and NACA high-speed wind tunnel tests of similar airfoils had shown that it would have a critical Mach number, unswept, of about 0.83 at zero lift. Sweep back and an aspect ratio of 3 were expected to raise this to "at least .99," which proved to be very optimistic. The aspect ratio was also expected to limit span-wise flow, which was known to reduce the effectiveness of the ailerons on a swept wing at a high angle of attack.

By contrast, the F-86 wing was

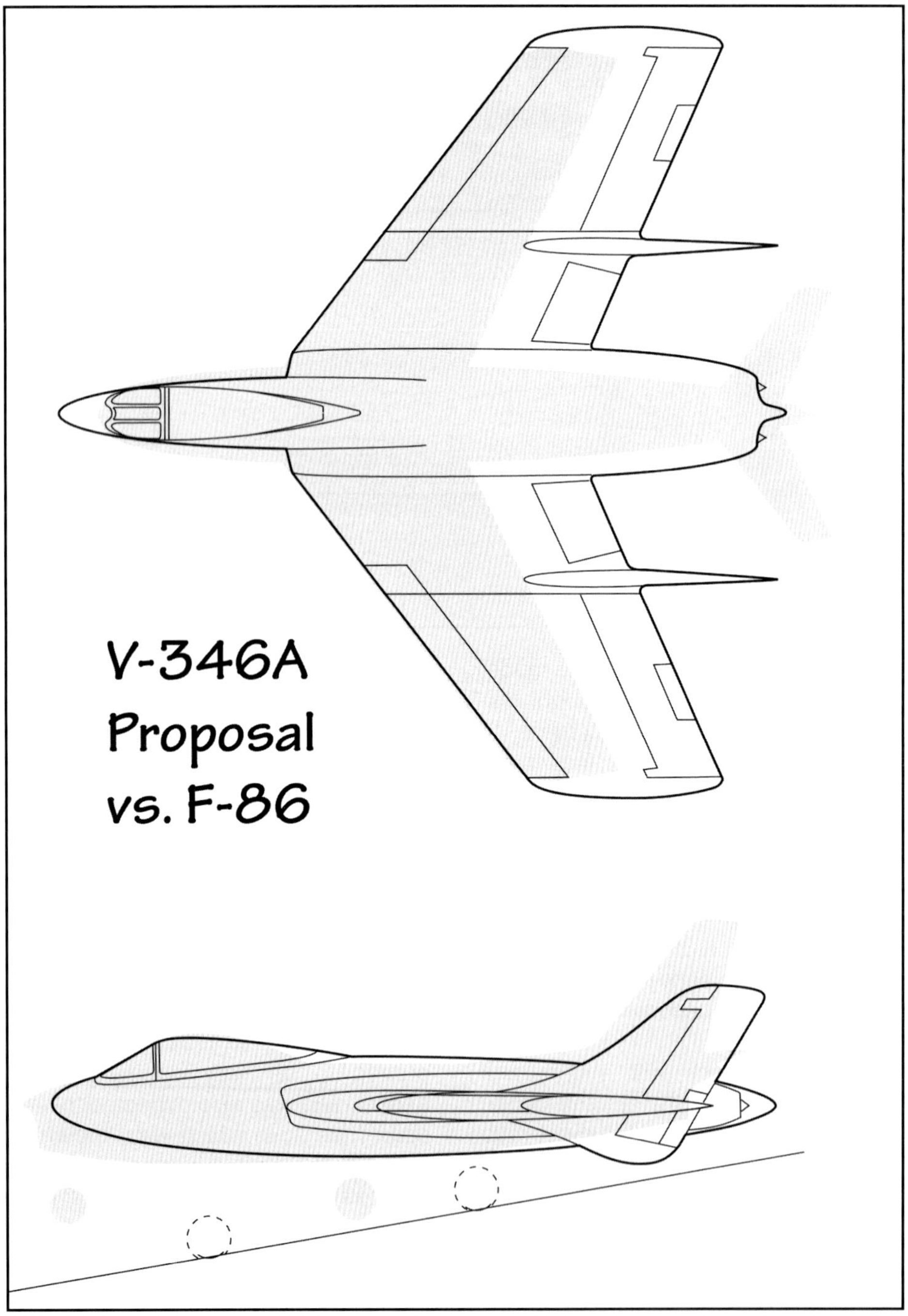

swept 35° and had an aspect ratio of 4.78 and a somewhat thinner airfoil: 9.5% thick at the root and only 8.5% at the tip. (Vought's conventional airplanes had a similar aspect ratio, 5.0 for the V-346B and 4.5 for the V-346D.) Vought's choice of wing thickness based on wind-tunnel test would prove to be a liability from a transonic performance standpoint. However, it did provide for more internal fuel capacity located on the center of gravity.

Since the V-346A had no horizontal tail, both pitch and roll control were accomplished with aileron-like control surfaces on the trailing edge of the outboard wing panels. Vought called them "ailavators," a combination of the words aileron and elevator, just as the ailavator combined the functions of those two flight controls. (Although "ailevator" might seem to be the correct portmanteau and it was spelled that way by others, Vought always spelled it with an a instead of an e.)

Vought initially developed the ailavator for Charles Zimmerman's very low aspect-ratio "Flying Pancake." The Navy funded an aerodynamic demonstrator, the V-173, and then placed a contract with Vought for a fighter version, the F5U.

At top, Vought had flight experience with combining the elevator and aileron surfaces from its V-173 Pancake program. In this case, the ailavators extended out from the side of the body. The surfaces on the trailing edge of the body were pitch trimmers. (Vought Heritage Center) At left, the V-346A was roughly the same size as the F-86 overall but the wing area was about twice that of the F-86, due to the need for a low stall speed and the inability to use flaps on a tailless design. Vought was clearly optimistic about the V-346A's empty weight. Relative to the F-86, it had a much bigger wing, more fuselage, larger internal fuel system, longer landing gear, powered vs. aerodynamically actuated slats, etc. However, Vought was estimating that it would weigh slightly less. (Author)

The Pancake did not have a wing, per se, so it did not have ailerons. Pitch and roll control were provided by a single surface, the so-called ailavator, on each side of the aft "fuselage." They moved together to pitch the airplane and differentially, to roll it. The test pilot had difficulty getting the V-173 to turn on its first flight. Vought subsequently modified the ailavators for more deflection and achieved satisfactory roll control. The F5U was never flown beyond "hops" along the runway.

The use of the same control surface for both pitch and roll dictated an analysis and wind tunnel test to insure that adequate control power was available in each axis in the worse-case flight conditions, e.g. roll control at low speed. In a conventional airplane, one aileron deflected full up and the other full down produces maximum roll rate. The elevators separately provide pitch control. Since both roll and pitch were provided by the same surfaces on a tailless airplane, Vought engineers had to take into account situations where the pitch-up control movement would limit the surface's down deflection that produced roll.

The twin fins avoided the potential blanking of a single fin at high angles of attack. Another benefit was that a pilot seat ejection at high speeds did not have to account for clearance of a centrally mounted fin. The base of the fin extended forward on the upper surface of the wing to act as a fence that would hinder the span-wise flow of air on a swept wing at high angles of attack, which reduced roll control power. The fin extended forward on the underside of the wing to house the main landing gear. The vertical fins and rudders suffered from the same lack of leverage for yaw control and stability as the ailavators relative to pitch control. The original size of the fins and rudders would prove to be incorrect.

The tailless concept imposed constraints on achieving a low takeoff and landing speed for carrier basing. Flaps could not be used to increase lift for takeoff and landing because when lowered, they caused an increase in nose-down pitching moment for which there was no horizontal tail to oppose. A high-lift airfoil, i.e. cambered, was also precluded for the same reason, pitching moment. Therefore the wing area had to be large relative to the weight of the airplane. This was a shortcoming from the standpoint of drag at high speeds but on the other hand, improved maneuverability at high altitudes. Leading edge slats, which increased lift at slow speeds with a minimal, if any, pitching moment, minimized the wing area required. To achieve the increased lift provided by the slats, however, the airplane had to be flown at an even higher angle of attack.

Above, the tailless Douglas F4D Skyray avoided the very high angle of attack on approach required by the V-346A by a combination of greater wing area, higher approach speed, smaller slats, and large "pitch trimmers" on the aft fuselage. (National Museum of Naval Aviation, Robert L. Lawson collection) Below, absent canards or other separate pitch-control surfaces, a flying wing could not utilize flaps due to the pitching moment that they created. Slats did provide an increase in lift, but not quite as much and at a significant increase in angle of attack on approach compared to an airplane using flaps for added lift. (Author)

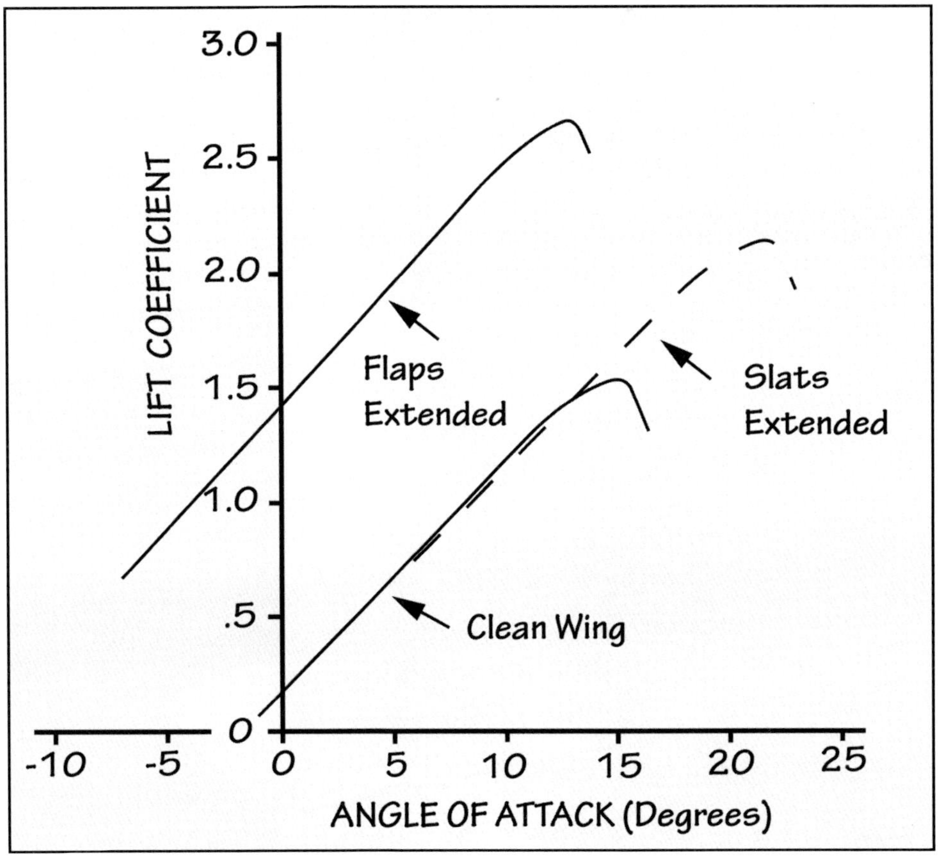

Above, the V-346A Artists Concept from the 15 April Vought proposal brochure. (National Archives, George Spangenberg papers) Below, in order to meet the range/endurance requirements of OS-105, the V-346 fuel system was comprised of seven fuel tanks with a total capacity of 1,035 gallons. The F-86 internal fuel capacity by comparison was 435 gallons. (National Archives, George Spangenberg papers)

The combination of wing loading, airfoil, and slat configuration that Vought settled on provided an acceptable stall speed but only at a very high body angle, which dictated a notably long nose landing gear. Douglas, by contrast, avoided an extreme nose-high takeoff and landing stance on its tailless F4D by a combination of a somewhat lower wing loading, far less proportion of the wing leading edge devoted to slats, large pitch trimmers on the aft fuselage, and a higher takeoff and landing speed.

Other V-346 Design Features

Before the advent of flight control computers, the location of the center of gravity (cg) was crucial for stability and control. Aft of the so-called neutral point, the airplane was unstable and difficult if not impossible to fly because the nose would not drop if the angle of attack increased due to turbulence. Instead, it would keep rising. The pilot had to balance the airplane as if it were sitting sideways on a log with a control ill suited to the task. If the cg were too far forward, the pilot no longer had enough control power at low speed to raise the nose. This made for longer takeoff runs and worse, an inability to flare for landing. It was also desirable that the airplane had a wide range of acceptable centers of gravity between too far aft and too far forward to accommodate changes in configuration and loading. The size and location of the horizontal tail was crucial in this regard. For one thing, it established how far aft the neutral point was relative to the aerodynamic center of the wing.

The tailless airplane presented two difficulties with respect to its center of gravity limits. First, because there was no horizontal stabilizer, the neutral point could not be aft of the aerodynamic center of the wing. Second, since the ailavators were located far closer to the center of gravity than the elevators would be on an airplane with a horizontal tail, they were less effective even when increased in size. The distance between the forward and aft cg limits was therefore half or less than that of a conventional design.

This limited cg range wasn't much of a limitation for a fighter, however, since most of the disposable load was fuel, which could be located on or near the center of gravity. It did, however, mean less tolerance for error in the as-built center of gravity relative to predicted center of gravity and dictated closer attention to fuel tank location and the timing of the use of the fuel in each tank. As it happened, rearrangement of large and small bits of the airplane was required during the design definition and detail design phases to move the actual center of gravity back into the desired relationship to the center of lift.

Although the fuel in the F7U would theoretically drain (in level flight) from the seven (subsequently eight) tanks evenly enough to keep the cg within the narrow acceptable range, a system of pumps and timers was incorporated to insure that happened.

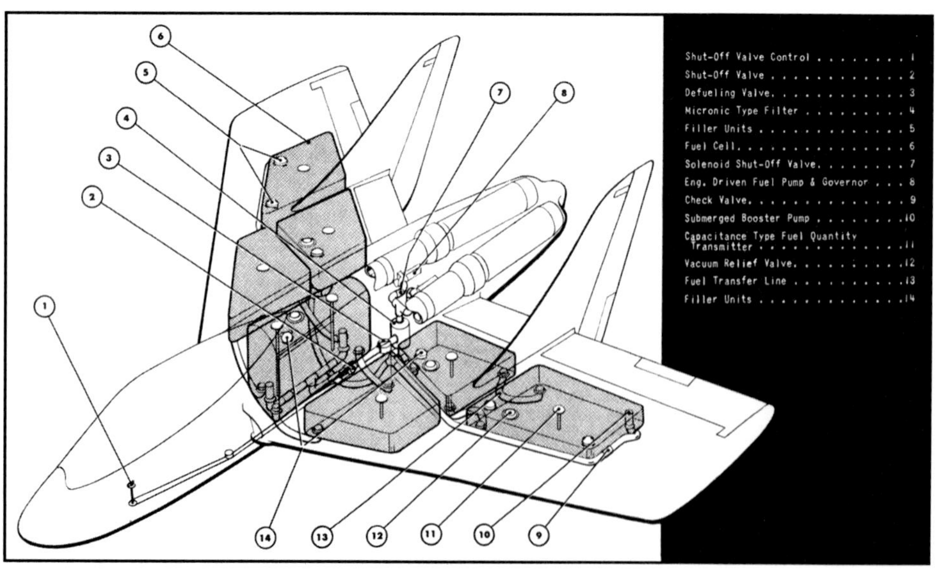

In order to minimize weight, Vought made maximum use of their patented Metalite and Fabrilite structure as well as magnesium, stainless steel, and high-strength aluminum alloy where appropriate.

- Metalite was a bonded (glued) sandwich of a balsa wood core and thin aluminum skins. It was relatively light and stiff compared to traditional aluminum structure. In this case, it was to be used for the fuselage skin around the cockpit and the wing skins above and below the fuel tanks. The cockpit skins not only provided the strength and stiffness required for pressurization, they maximized the internal volume due to the elimination of stringers and the minimization of frames. They also provided a modicum of thermal insulation and sound deadening. The thin Metalite wing skins similarly allowed for a maximum volume of fuel.

- The skin on the nose and the upper fin and rudder were made out of Vought's Fabrilite (basically Metallite with Fiberglas instead of aluminum skins) to allow for the transmissivity of signals to and from antennas located there. The stiffness of the skin also allowed for a minimum number of ribs and spars.

- Magnesium was even lighter than aluminum but also more expensive and not suitable for some parts because of its poor resistance to corrosion. On the V-346 it was to be limited to the ailavators and slats, which were readily replaceable and some secondary structure where it could also be inspected and replaced if required due to excessive corrosion.

- The aft fuselage was constructed largely from stainless steel in order to withstand the heat generated by the afterburners. (Its utilization on the F7U-1s was readily apparent because it remained unpainted.)

- For highly-loaded parts like the landing gear and some of the fuselage structure, 75ST aluminum alloy was used instead of 24ST for the same strength at less weight, albeit at higher cost and notch sensitivity from a fatigue standpoint.

Hydraulically-powered flight control systems were rare in propeller-driven fighters but becoming standard in jets because of the control loads in high-speed flight. To minimize weight, engineers initially attempted to make do with a single hydraulic system, operating both utility (landing gear, flap, slat) actuators as well as the flight control surfaces. Having only one hydraulic system required that the flight controls receive priority if total demand at any point exceeded supply at any point, for instance if an engine or a pump had failed, reducing the flow available. It also meant that the pilot had to be provided with an unboosted control capability in the event that hydraulic pressure was lost due to pump failure or fluid loss.

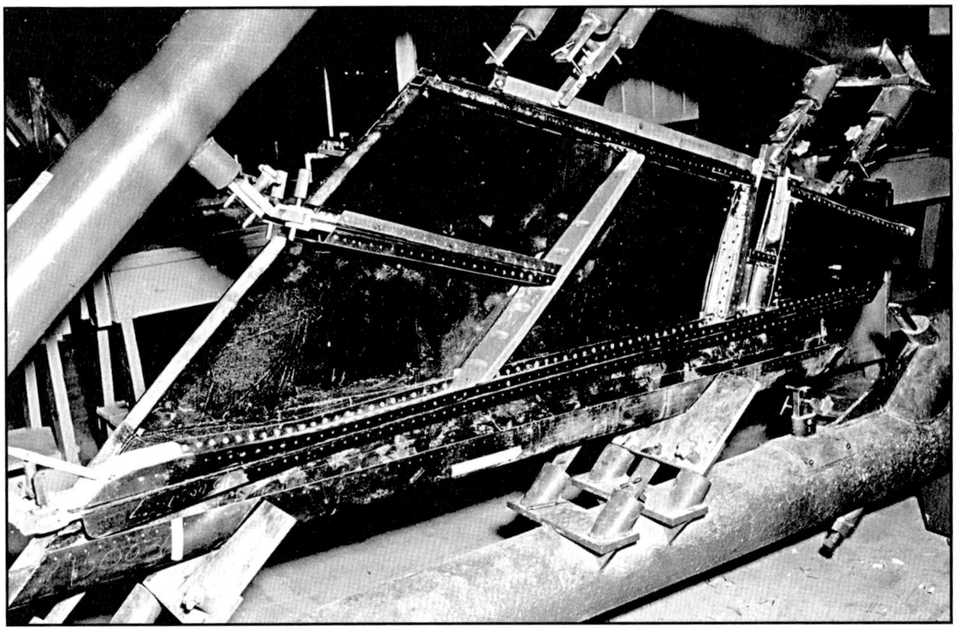

Above, Vought innovations included light and stiff skin panels consisting of a balsawood core sheathed on both sides by either a thin sheet of metal (Metalite) or fiberglass (Fabrilite). Lying on the table in the foreground are balsa-wood cores for Metalite wing panels. (Vought Heritage Center) Below, the strength and stiffness properties of the Metalite and Fabrilite panels minimized the requirement for spars and ribs and eliminated stringers in structure like the F7U-1 vertical fin shown here. In the case of the fin, they also provided transmissivity for a fully-enclosed radio antenna, which reduced drag. (Vought Heritage Center)

The proposal specification didn't provide much detail on the control system: "The contractor shall furnish and install a hydraulic power boost system for ailavators and rudders. In case of failure of the hydraulic power boost system, it shall be possible to control the airplane by direct manual control aided by trim tabs."

In accordance with that practice, the F7U-1 therefore had a single hydraulic system although two pumps were provided, one on each engine. Vought used a nominal 3,000-psi hydraulic system instead of the then standard 1,500-psi. (Some utility circuits were metered down to 1,500 psi in order to allow the use of existing components.) The higher pressure allowed the use of smaller (and therefore lighter) hydraulic lines and required less hydraulic fluid.

The proposed empty weight was 9,075 lbs. (Even without the benefit of hindsight, this seems optimistic since it was 600 lbs less than the F4U-5 Corsair's empty weight.) To that was added the pilot, mission equipment, 5,826 lbs of fuel, 1,000 lb of guns and ammunition, etc., for a gross weight of 16,797 lbs.

BuAer provided Vought with a letter of intent on 25 June 1946 for the design, manufacture, and test of three XF7U-1s based on the V-346A proposal. At that point, Vought's first jet airplane, the XF6U Pirate, had not flown and would not for another three months. Moreover, it was a small, single-engine, straight-wing airplane, not very far removed aerodynamically from propeller-driven fighters and as previously mentioned, did not have either an ejection seat or cockpit pressurization. Although the production Pirates were to incorporate afterburners, the first XF6U flight with one was even farther off in the future. The incorporation of an ejection seat and cockpit pressurization would be implemented with the production F6Us.

As with almost all government aircraft procurements, hardware common with other airplanes of the same type were provided as government furnished equipment (GFE) to reduce procurement and operating cost. These included the engines (and in this case, separate afterburners), armament, gun sight, generators, bat-

Below, the V-346A proposal complied fully with OS-105 requirements including avionics and other mission equipment. The inlet duct door was a feature of early jet airplanes to keep an engine that had been shut down in-flight to increase endurance from windmilling, which could cause bearing damage. (National Archives, George Spangenberg papers)

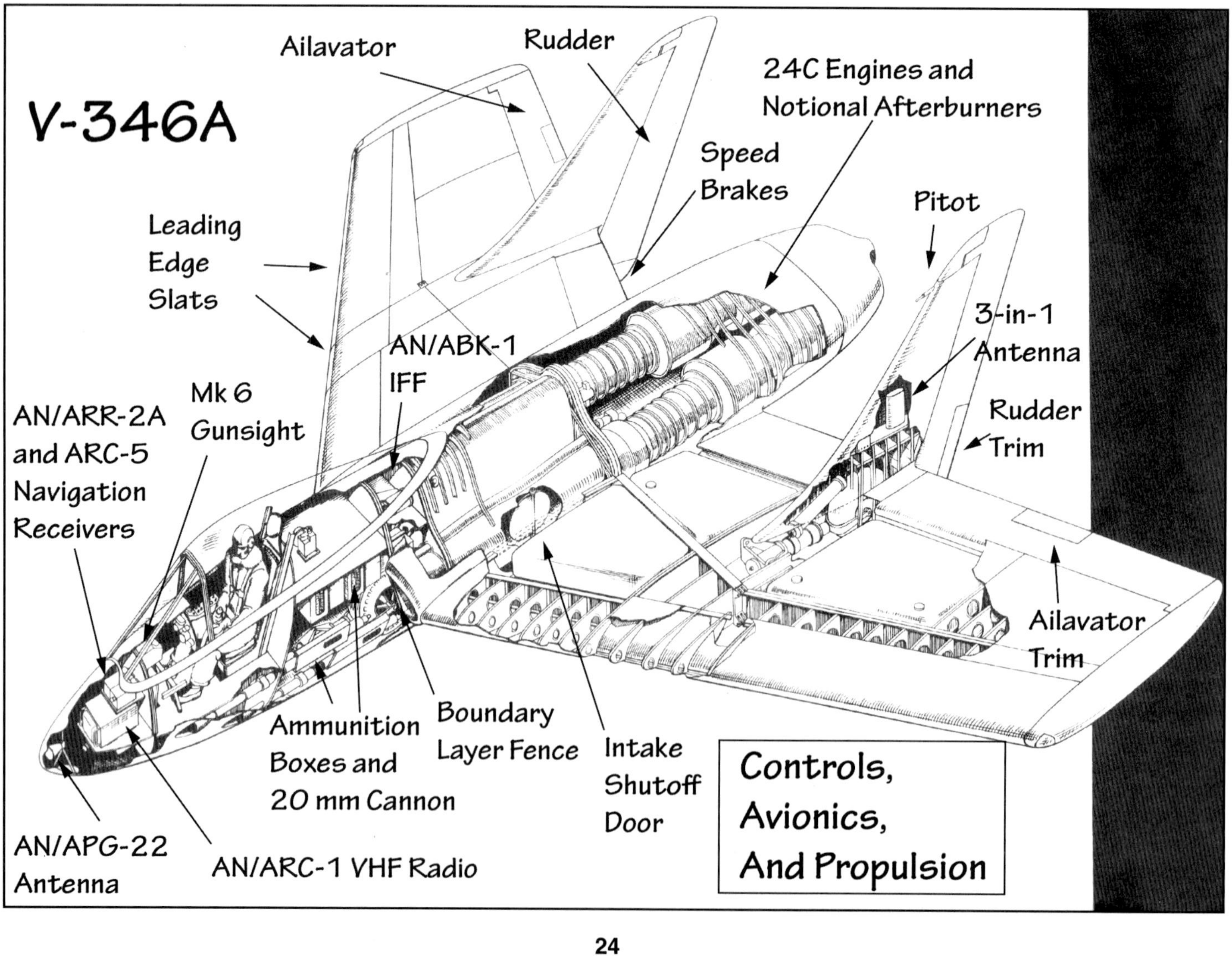

At right, the little Vought F6U-1 was powered by a single Westinghouse 24C engine and a Solar afterburner. The V-346A was to be powered by two of the same engines/afterburners. Its empty weight was projected to be only about 24% greater than the F6U's. While the armament, avionics, and cockpit furnishings were the same, the weight of the additional engine/afterburner alone was more than the difference in empty weight even before the V-346A's larger wing, fuel system, and landing gear were considered. (Author)

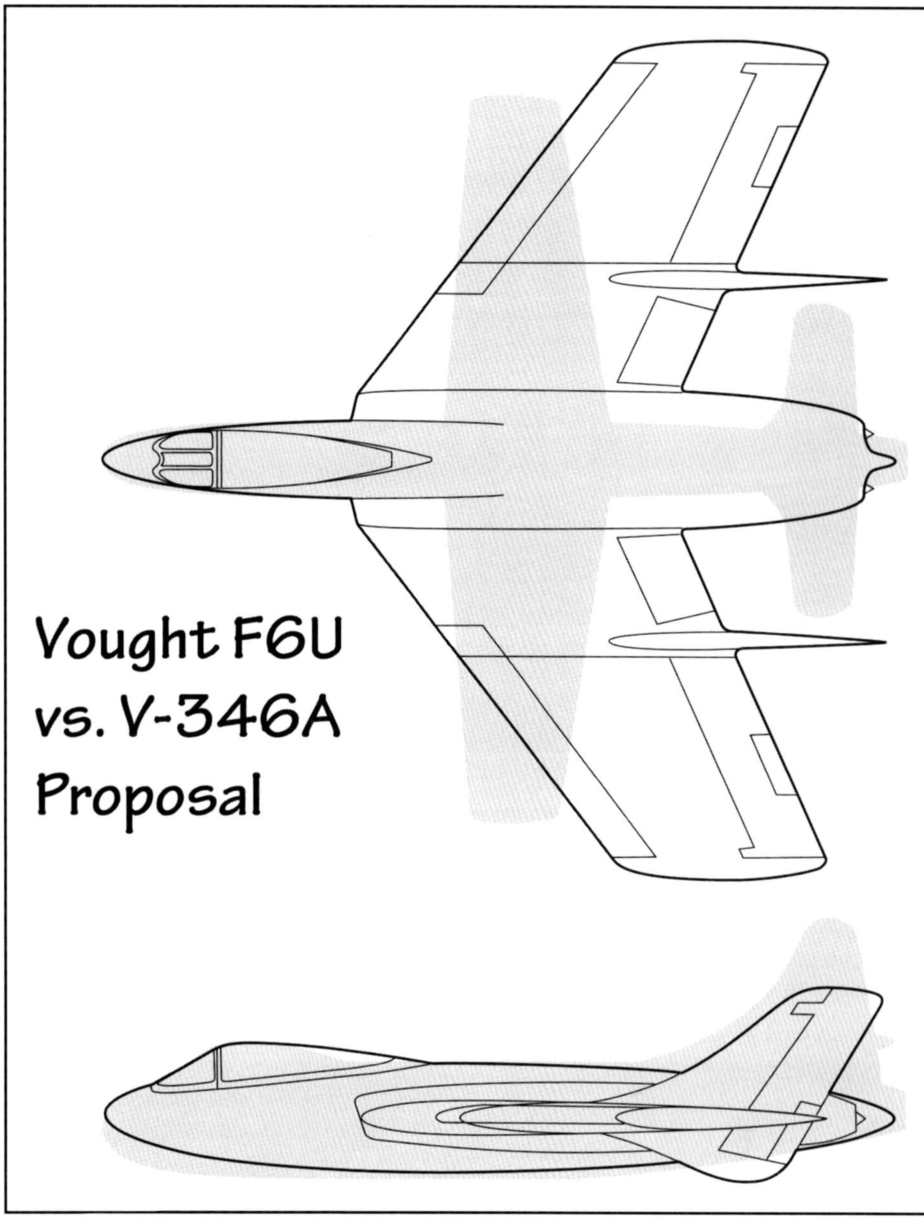

teries, instruments, radios, etc.

SD-430, the Detail Specification for the XF7U-1, was dated 8 July 1946. During the refinement of the specification in negotiations with the Bureau of Aeronautics, the proposed empty weight increased "considerably" according to the 1 August 1946 BuAer Aircraft Division Monthly Progress Report. Reasons cited were "(A)dditions in afterburning equipment, firewalls for engine units, fire detectors and extinguishers, gun sight weight, fuel tank and ammunition feeding change, and wing redesign to shift C.G. forward and lessen range of C.G. movement." The guaranteed empty weight was 9,711 lbs, an increase of 637 lbs from the V-346A proposal.

At right, the tailhook location and configuration had already been changed from the original V-346 study and would continue to change as the design evolved from mockup through production. (Spangenberg papers)

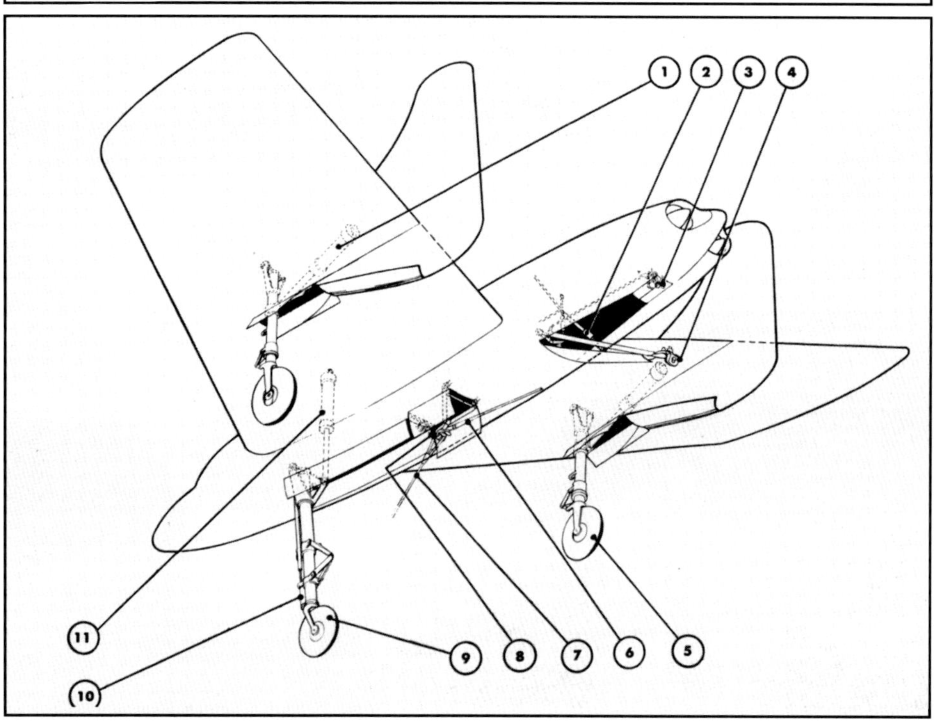

1.) Self-locking Hydaulic Actuating Strut & Drag Brace
2.) Dash Pot
3.) Arresting Hook Bumper
4.) Replaceable Head Type Arresting Hook
5.) 30 X 7.7 Extra High Pressure Non-Skid Tire
6.) Catapult Holdback & Release Unit
7.) Launching Pendant
8.) Retractable Catapult Hook
9.) 24 X 7.7 Extra High Pressure Non-Skid Tread Tire
10.) Centering Mechanism & Hydraulic Shimmy Damper
11.) Self-Locking Hydraulic Actuating Strut & Drag Brace

Above, the F7U-1 mockup's instrument panel was the latest in a lengthy gestation period that featured many changes. (USN, National Archives) Below, as usual, the proposed cockpit layout was only a starting point for an ongoing series of changes resulting from design and mockup reviews and flight evaluations. The large unlabeled lever on the left-hand console is probably the hydraulic boost shutoff but it may have been to provide one throttle for two engines. (National Archives George Spangenberg papers)

Performance guarantees were:
Maximum speed in level flight at military rated thrust: 538 knots (619 mph)

Maximum rate of climb at military rated thrust: 6,450 feet per minute

Stall speed, power off, 25% fuel: 78 knots (90 mph)

Takeoff rate of climb at military rated thrust, gear down: 830 feet per minute

Up to that point, BuAer had managed to minimize the gross weight growth in spite of changes requested by the Navy or GFE weight increases. The most significant was an increase in the engine and afterburner weight, which required Vought design changes to rebalance the aircraft. To offset these weight increases, BuAer had agreed to reduce the pilot's armor protection, the takeoff fuel from 1035 gallons to 971 (with the proviso that space should be reserved to reinstate the capacity) and the ammunition from 1000 rounds to 800 (again, with space for the original capacity to be retained). Because of the fuel quantity reduction, the radius of action was reduced by about 40 nautical miles.

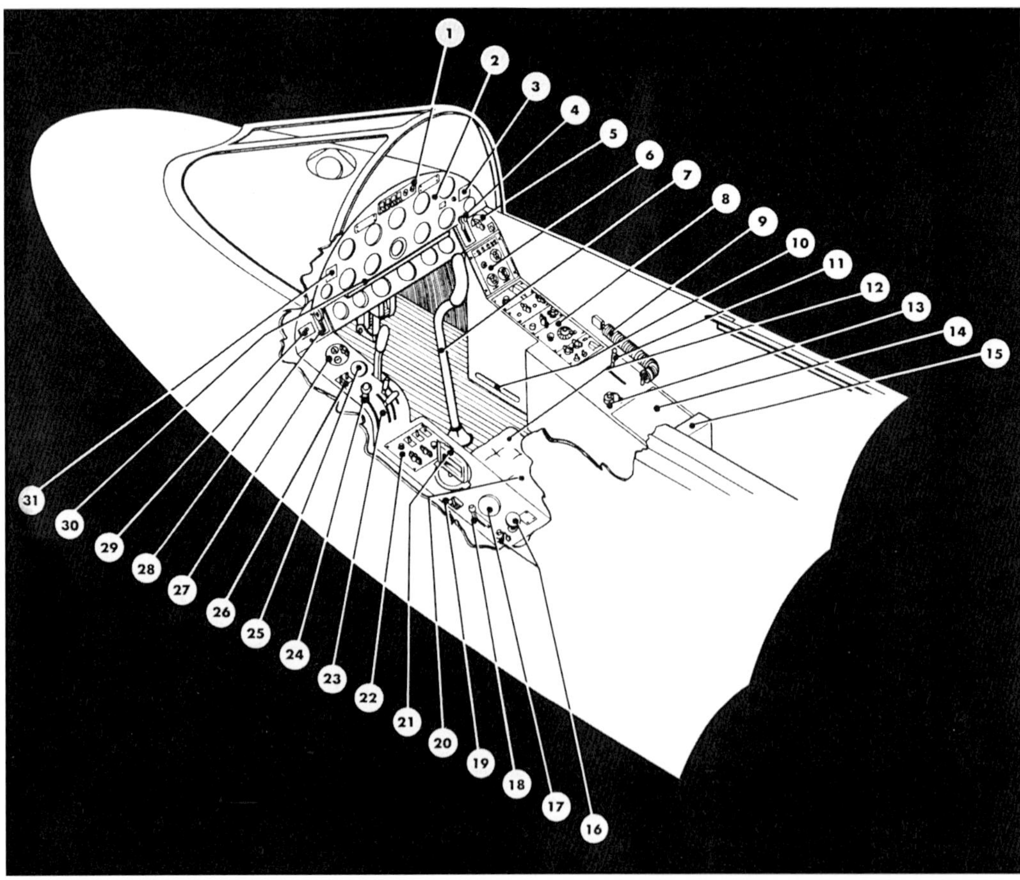

Mockup Review

The review of the detailed, full-scale mockup was held at Stratford, Connecticut, on 15 to 18 October 1946. Vought's first jet, the F6U, had flown for the first time only two weeks earlier. In addition to the airplane mockup, there was a separate, more detailed cockpit mockup that was available for a preliminary evaluation on 7 October.

There were several notable changes from the proposal:

I. The wing and engine inlets were moved aft about 10 inches
II. The landing gear was lengthened to provide more fin clearance and the fins moved aft relative to the wing to provided a longer landing gear well.
III. The single fuselage tank was replaced by two tanks and the ammunition was moved aft to provide less movement of the center of gravity.
IV. The engine was moved forward as a result of revised definition of the afterburner length and weight.
V. The canopy now had a separate clear hood, reducing the rearward vision, and a short afterbody, with the hood sliding aft over the afterbody.
VI. The tailhook was again attached at the aft end of the fuselage, but as a Grumman stinger-type hook that extended aft out of the fuselage and then dropped into position. It also incorporated a pendant-shedding capability.

The canopy change from the proposal apparently stemmed from a BuAer engineer's concern about the strength of the canopy proposed originally that had excellent rearward visibility. The tailhook installation change was only the latest in a series and not the last. In the February V-346 design study, the tailhook was attached at the aft end of the fuselage and pivoted upwards to stow backwards (hook point forward and throat upside down) in a compartment on the top of the aft fuselage. For the V-346A proposal, it had been relocated to the bottom of the fuselage in a more conventional arrangement; however, this was found to create an excessive nose-down pitching moment on arrestment.

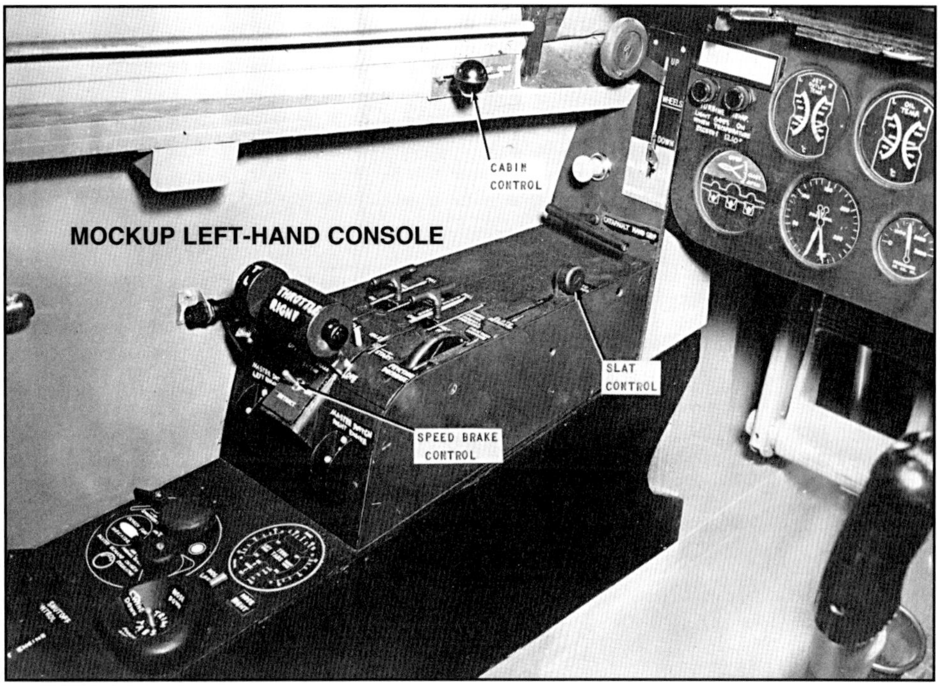

MOCKUP LEFT-HAND CONSOLE

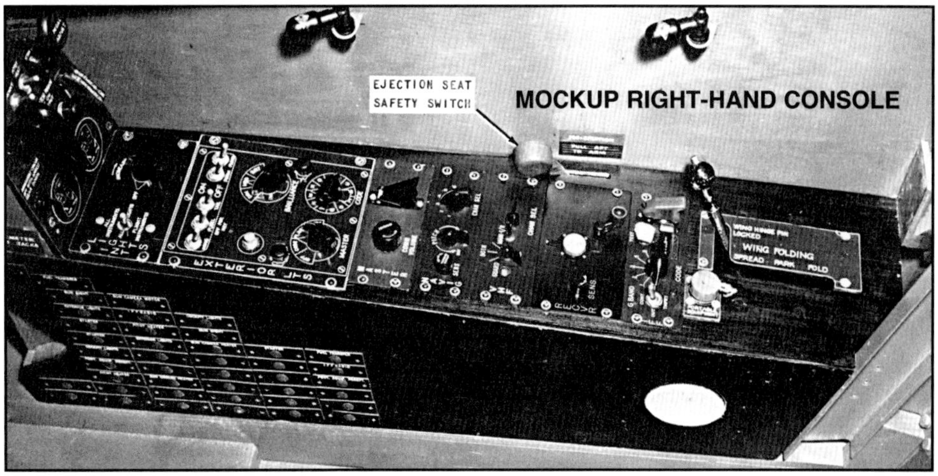

MOCKUP RIGHT-HAND CONSOLE

MOCKUP RIGHT-HAND CONSOLE

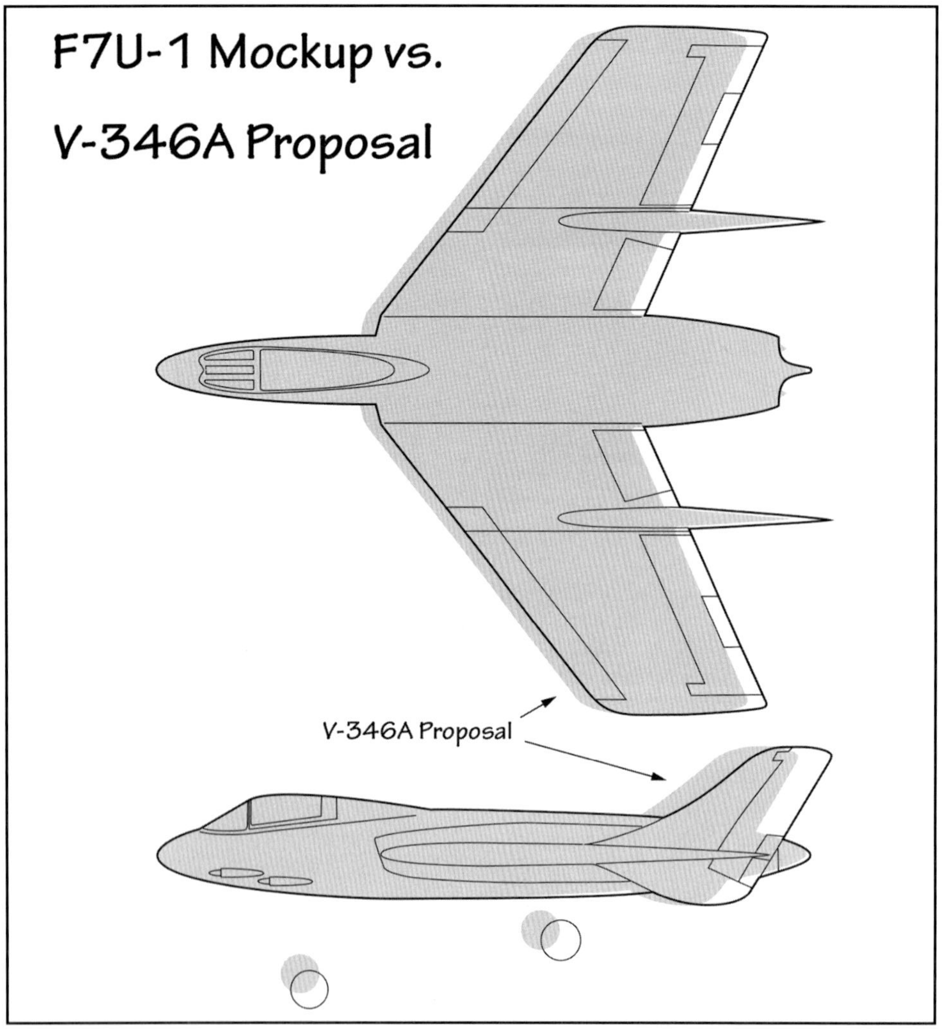

At left, two-view comparison of V-346A and Proposal: In order to properly position the center of gravity, the wing had to be moved aft relative to the fuselage in addition to various internal rearrangements. The fins were relocated still farther aft to help the rebalance and provide the longer wheel well needed for the longer main landing gear. (Author)

	Estimate	OS-105
Vmax, sea level and combat power (mph)	663	650
Vmax, 8,000 feet and military power (mph)	617	600
Vstall, takeoff gross weight (mph)	105.1	105
Vstall, landing weight with 25% fuel (mph)	90.3	90
Takeoff distance, no wind (ft)		1682
Takeoff distance, 25 kt wind (ft)		970
Rate-of-climb at sea level*, combat power (fpm)	9700	10,000
Rate-of-climb at sea level*, military power (fpm)	5,420	6,500
Rate-of-climb at takeoff (fpm)	830	500
Service ceiling, military power (fpm)	44,000	40,000

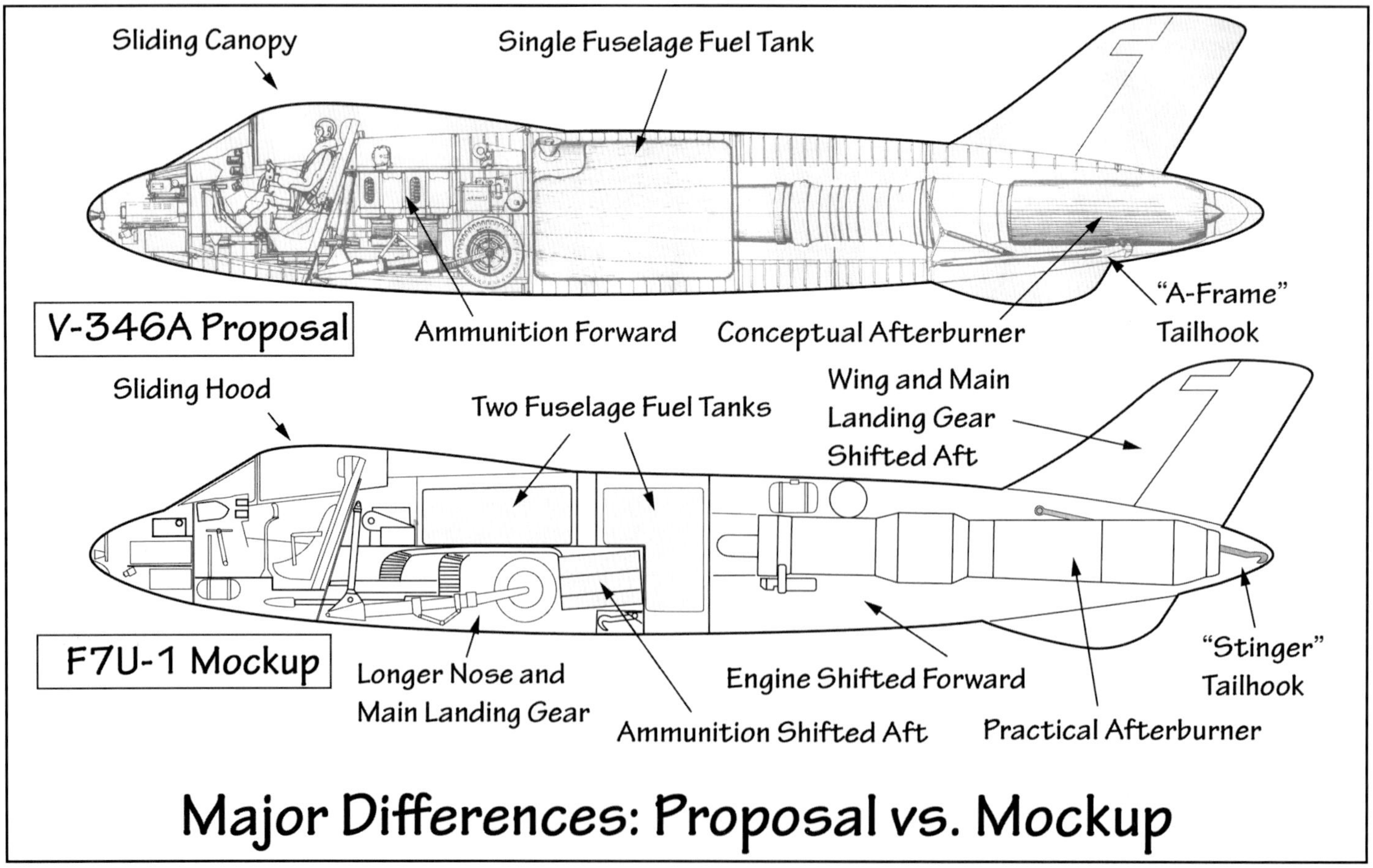

*Combat gross weight (not to be confused with combat power) was the weight at mid-mission, calculated with 60% of full internal fuel, or 14,461 lbs. At combat weight and combat power, the airplane's rate of climb was estimated to be 11,290 fpm at sea level; at military power, it was to be 6,390 fpm.

Vought provided the attendees with the following performance projection. All estimates were at a takeoff gross weight of 16,795 lbs except for the stall speed at landing weight. Military power was defined as the maximum non-afterburning thrust. Combat power was with the afterburner operating.

There were 67 comments/suggestions/complaints recorded by mockup board attendees. According to the formal mockup board report, "No items of major controversy arose." Due to weight and other considerations, the Mock-up Board accepted few of the suggested changes. The resulting empty weight would have actually gone down by four lbs but for a 23-lb error made by BuAer in conveying the weight of the Mk 7 Aircraft Fire Control System to Vought.

One change that was accepted was the configuration and operation of the nose gear doors. They were originally left open after the gear was extended. No protective cover was provided for the armament and other equipment located in the nose wheel well. A fabric cover was suggested and mocked up but it was determined to be inadequate. The Board opined that either a sheet-metal cover be provided on the sides of the wheel well or the nose wheel well doors be closed after the nose landing gear

Above, the F7U-1 mockup was very complete and detailed except for the engine and afterburner installation. For example, the landing gear could be retracted to depict the flight configuration. However, there were some changes yet to come like the chute on each side of the lower forward fuselage that funneled the guns' cartridges and links aft under the engine inlet. (USN, National Archives) Below, in addition to the retractable landing gear, the F7U-1 mockup featured folding wings. (USN, National Archives)

At left, there were many detail changes to the F7U configuration compared to the proposal. In addition to those shown here, the main wing spar was relocated (and swept aft on the inboard wing panel as it was on the outboard) and the capacity of the fuel tanks in the wing modified accordingly. (Author)At right, the large leading edge slats could be extended and retracted. (USN, National Archives)

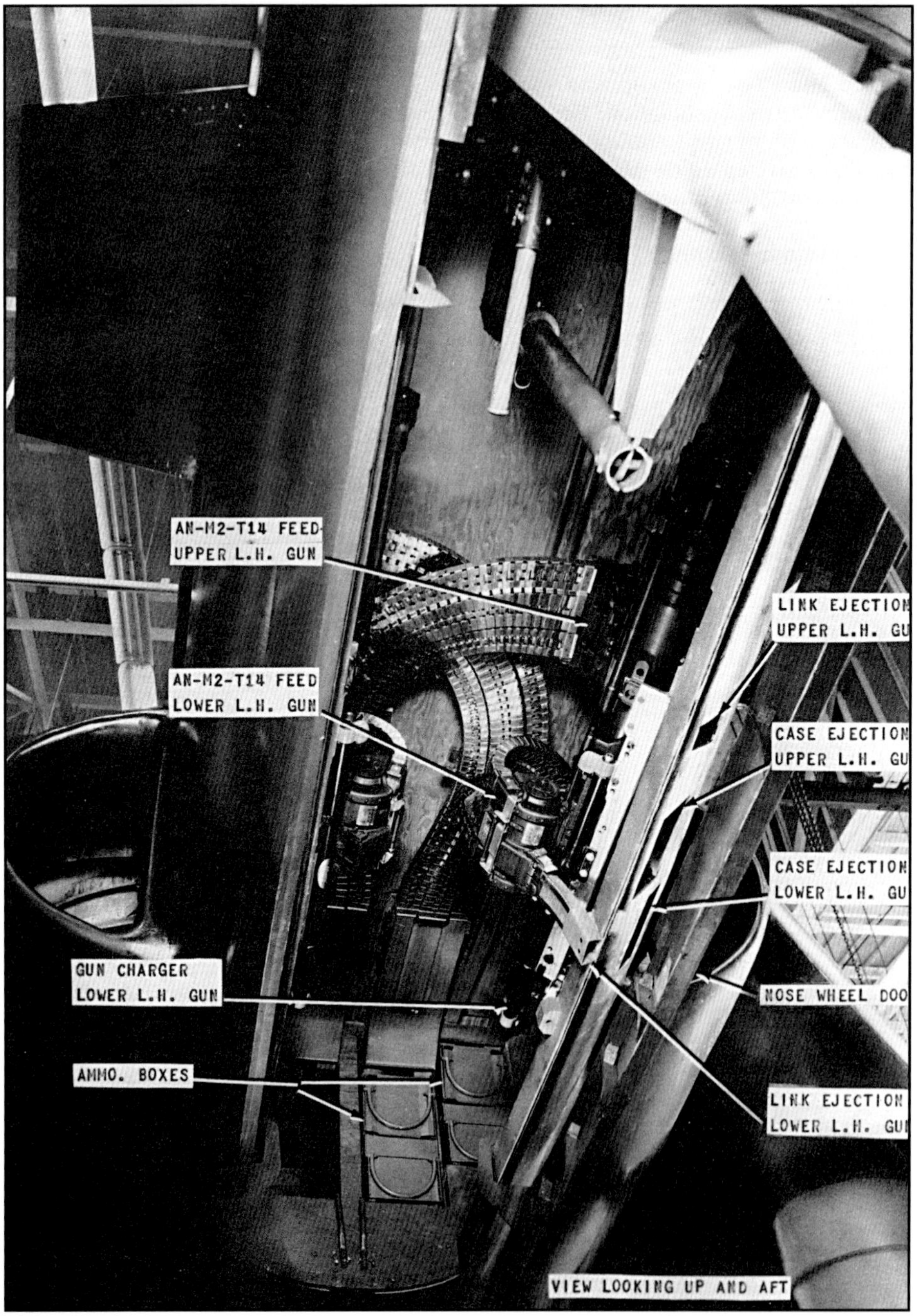

had been extended. The agreed-to solution was to add a small front-facing door so the nose wheel well doors were open only for nose landing gear extension/retraction and as required for maintenance and replenishing the ammunition.

The mockup was inspected again on 23 December 1946. The purpose was to evaluate some of the changes that had been approved at the original review, principally the gun sight, ejection seat, and cockpit-access control installations. Vought had also requested approval of a change from the two-piece canopy on the mockup back to a one-piece canopy.

The major change involved the gun sight installation, because the original was determined to require too large a head movement from the normal seating eye position to the gunsight eye position. As a result of the requested Vought design study, the windshield was moved aft 1.5 inches. This in turn required that the pilot's seat be automatically lowered to the full-down position for ejection (when it was lowered, the seat also moved aft) in order to provide adequate clearance between the pilot's knees and the instrument panel.

At the time of the mockup review, the ejection seat was to be actuated by the arm rests like the Army Air Forces-specified approach whereas the Navy was leaning toward the Martin-Baker concept of having the pilot pull down a curtain over his face to initiate ejection. For the time being, the Board elected not to make the change, pending the results of NAMC (Naval Air Material Command) ejec-

At top left, the large wing-mounted speed brakes were to be extended and retracted with a rotary actuator. (USN, National Archives) At left, the nose landing gear wheel well provided access to the 20 mm cannon and ammunition boxes so external doors were no longer required as depicted in the V-346A proposal. (USN, National Archives)

tion seat test and evaluation then in progress. The final configuration utilized the face-curtain actuation.

Pulling the face curtain was actually the third step in the ejection process. The first was to position his feet in the stirrups and put his head against the headrest. After that, due

At right, the tailhook was another installation that would undergo scrutiny and change. On the mockup and prototypes, it would extend like a stinger. However, it was initially to be housed behind clamshell doors. The slot in the top of the hook point was for a "cable shedder" that would insure that the arresting cable would drop from the hook when it was raised after landing. (USN) Below right, the mockup main gear and wing fold point. (USN) Below, the nose landing gear was configured with a two-sided yoke and a side-mounted shimmy damper on the mockup. It would undergo extensive changes before and after first flight. (USN)

to concerns about the seat firing without the pilot having initiated ejection, he pushed a lever that placed the seat in its lowest position, decompressed the cabin, opened and released the canopy, and armed the seat. When all of those steps had been accomplished, a red light on the instrument panel illuminated to indicate to him that it was safe to eject by pulling the windscreen located at the top of the headrest down over his face.

After considerable discussion, a "Bypass" position was added to the canopy control in the cockpit. The external canopy release was moved next to the aft fuselage step on the left-hand side of the airplane. Its function was now to reset the cockpit control to bypass so that the canopy could be opened manually. (The prior plan was to open and close the canopy externally by means of a toggle switch and hydraulic hand pump located in the nose wheel well; if the area was not accessible after a crash, a bypass valve located on the right side of the fuselage allowed the canopy to open manually.)

In conjunction with this change, the Mockup Board approved the canopy change back to a conventional one-piece sliding design after it was determined that it was lighter, simpler in operation, easier to open and close manually, and could also be opened at high speed. However, the reduced glass area was retained in order to insure the structural integrity of the design.

At top left, most of the avionics were grouped in the removable nose cone. The AN/APG-22 radar antenna is not shown here because it was mounted in the nose cone itself. (U.S. Navy, National Archives) At left, the Navy became enamored early on with the Martin-Baker vision for ejection seat design. It bought a test tower from Martin-Baker in 1946 and installed it at the Philadelphia Navy Yard. Here, Mr. James Martin himself is insuring that the test subject is properly strapped into his seat. (Martin-Baker)

Aerodynamic Design Support Tests

Vought proposed and accomplished the standard wind tunnel testing for lift, drag, and pitching moment. However, some additional testing was accomplished that related to its unusual configuration and transonic capability. As it happened, BuAer and NACA Langley in Virginia had a close working relationship and much of this testing was done there.

Power-off Stability and Control

One was the "flight" of a 1/10 scale low-speed model in the NACA Langley 12-ft free-flight tunnel between 13 May and 14 August 1947 for evaluation of power-off stability and control. The dynamic model (meaning its dimensional and mass characteristics were scaled the same and its airfoil was selected to have the same maximum lift coefficient at low Reynolds numbers as the full-size airplane) was flown in the tunnel via control of the speed of the tunnel air and its angle to the model for longitudinal and vertical control. A "pilot" remotely controlled the model's rudders and ailavators for lateral and yaw movement. Before a run, the leading-edge slats would be positioned either retracted or extended.

The longitudinal and lateral stability characteristics were satisfactory over the lift range up to the stall. With the slats retracted, the stall was fairly gentle but the model rolled off out of control. With the slats extended, control could be maintained at the stall so that the wings could be kept level even as the model dropped. Vought used a movie of this test to maintain interest and support at BuAer after the mockup review and before first flight.

A 0.08-scale high-speed model was tested in NACA Langley 7x10 foot High Speed Wing Tunnel over a Mach number range of 0.40 to 0.91 in 1947 to provide lift, drag, and pitching moment data of the basic airplane with undeflected controls. To avoid interference from the usual strut mounts used, the model was support-

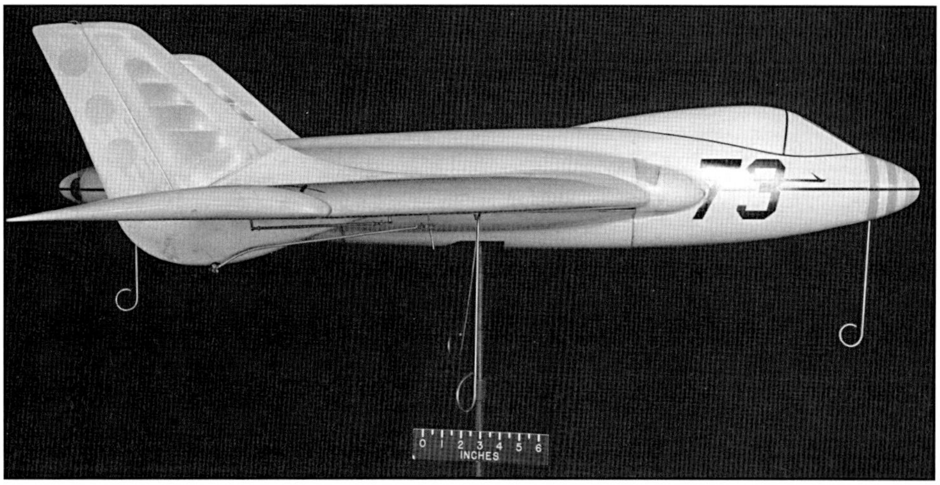

Above, this large (1/10 scale) "free-flight" model was tested in the NACA Langley 12-ft low-speed wind tunnel in 1947 to determine power-off stability and control derivatives. It was lightly built so it was dynamically equivalent to the actual airplane and had remotely-controlled rudders and ailavators. (NASA LMAL 52458) Below, the answer to the need for increased lift was established to be extension of the leading edge slats all the way inboard to the engine nacelle. On this 1/10 scale model, the inboard-slat chord was somewhat greater than proved to be practical on actual airplane due to the need to retain the fuel capacity of the forward inboard fuel tanks. Movies of this testing were used by Vought to maintain customer confidence in the F7U configuration before its first flight. (NASA LMAL 52216) Bottom, basic lift, drag, and pitching moment data were obtained from a .08 scale model tested in the NACA Langley 7x10-ft High Speed Wind Tunnel at speeds up to Mach 0.91. (NASA)

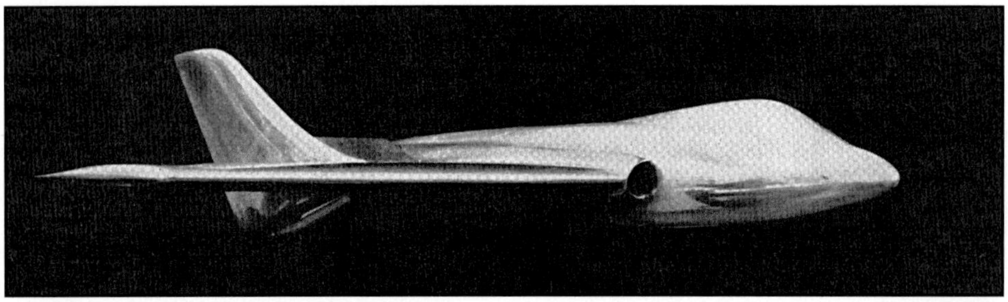

Above, the F7U-1 wing-flow model was about a foot long and carved from steel. (NASA) Below, taking advantage of the fact that the local airflow over the wing of a P-51 Mustang in a dive was transonic to slightly supersonic, a "wing-flow" test method was developed to obtain data at those airspeeds. (NASA EL-2000-00367)

ed in the tunnel by a sting extending from the rear of the fuselage to a vertical strut located well behind the model. The strut was mounted on a balance system to measure loads and moments. The model could be positioned before the test at the desired angle of attack.

Wing-Flow Testing

Until the successful demonstration in late 1950 of the slotted-throat wind tunnel, data on the speed range between approximately Mach 0.95 and 1.1 could not be provided by wind tunnel test. Even then, it was only valid at relatively low Reynolds numbers. Since the F7U was projected to reach a dive speed of Mach 1.1, NACA Langley suggested in June 1946 that a 0.026-scale "wing-flow" model also be tested to cover the speed range around Mach 1. This drag and pitching moment measurement tool took advantage of the fact that airflow over a portion of the upper surface of a P-51's wing was supersonic in a dive. The model consisted of one half of the airplane—including the fuselage, air inlet duct, vertical tail, dive brake, and elevator—mounted on a steel plate that could be rotated through small angles to present the model at angles of attack from -3 degrees to 14 degrees at Mach numbers from 0.65 to 1.08. The ailavators could be deflected to a desired angle before takeoff. The model was "made essentially of solid steel" because of the high loads expected.

Test runs were made at each of five ailavator settings in dives from medium and low altitudes and in low-altitude, high-speed, level flight.

The results of the tests indicated no adverse effects of compressibility up to Mach 0.85 at low normal-force coefficients and small ailavator deflections. At that point, the neutral point for lift moved rearward from 25% to about 42% at Mach 1.05. Loss of ailavator effectiveness or actual reversal occurred between Mach 0.93 and 1.0. This was typical of transonic flight and meant that a nose-down trim change of increasing magnitude would occur as the airplane accelerated toward Mach 1 and then a more abrupt change of ailavator position, first up and then down, would be required to maintain pitch angle.

Spin Testing

Spin testing of a .057-scale model in the NACA Langley 20-feet free-spinning tunnel was accomplished to determine the spin and recovery characteristics and also the tumbling tendencies of the XF7U-1. The model was ballasted with lead weights so that it was dynamically similar to the actual airplane at an altitude of 15,000 feet. A remote-control mechanism was installed in the model to actuate the control surfaces for spin recovery. A wing-tip mounted spin recovery parachute installation was also evaluated. For the tumble tests, a remotely controlled tumble-recovery parachute was incorporated. In both series, a small dummy pilot with parachute was released to determine if emergency egress from a spin or tumble was feasible without benefit of the ejection seat.

No untoward characteristics were identified, although NACA noted that spin-tunnel results were not always in complete agreement with full-scale results. About 80% of the predictions were considered accurate, with the remainder being optimistic or pessimistic in equal measure. The recommended recovery technique was to hold the stick full back and laterally neutral while full rudder was applied against the spin. After a half turn but no more than one turn later, a brisk movement of the stick forward was to

be made, keeping it laterally neutral. The pilot was cautioned to avoid move the stick forward too early and also to avoid pulling out of the ensuing dive too quickly.

Since there were no full-scale tumbling events for comparison purposes, the fact that the model did not continue to tumble even when falling back from a vertical launch or launched with an initial pitching rotation was viewed as good news. It could be made to tumble at an aft center of gravity with the ailavators deflected full up but only if given a positive external pitching moment; it recovered immediately with downward deflection of the ailavators. However, the accelerations at the pilot's head in the tumble were considered to be high and immediate recovery was recommended.

For purposes of this test, a worse case scenario for pilot escape was used in that he was released and not ejected. The dummy pilot in the tumbling airplane was observed to clear all parts of the model by a large margin.

It Takes All the Running You Can Do To Stay In Place

In the proposal calculations, Vought had estimated that the V-346A wing had a maximum lift coefficient of 1.2. This turned out to be optimistic on further examination of the pitch attitude attainable with the ailavators. Adding wing area would be a last resort because of the need for compactness on a carrier.

One solution was the use of auxiliary trim devices to provide more angle of attack. The ailavators actually were a handicap in this regard because they reduced lift initially when used as elevators and when deflected fully, adversely affected roll control.

Vought accomplished wind tunnel test of a 1/7th scale model in an MIT wind tunnel in September 1946 and in one at the David Taylor Model Basin in November. The objective was to evaluate auxiliary trim devices that would increase lift for takeoff and landing without compromising longitudinal control, lateral control, or wing-stall characteristics.

Above, the wing-flow model was mounted on a test fixture installed in the gun bays of the P-51 test aircraft. It was pivoted in flight to obtain data at desired angles of attack. The ailavator angle could be set before flight. (NASA)

In this series of wind tunnel tests, Vought evaluated auxiliary control surfaces to provide the requisite nose up trim at high angles of attack. These included flaps that extended outward from near the top of the vertical fins, split rudders, a so-called

Below, wind-tunnel spin testing was not only accomplished to determine the spin characteristics of the F7U-1, but also to verify that the spin recovery parachute system to be used during the airplane spin-test program would function as designed. (NASA)

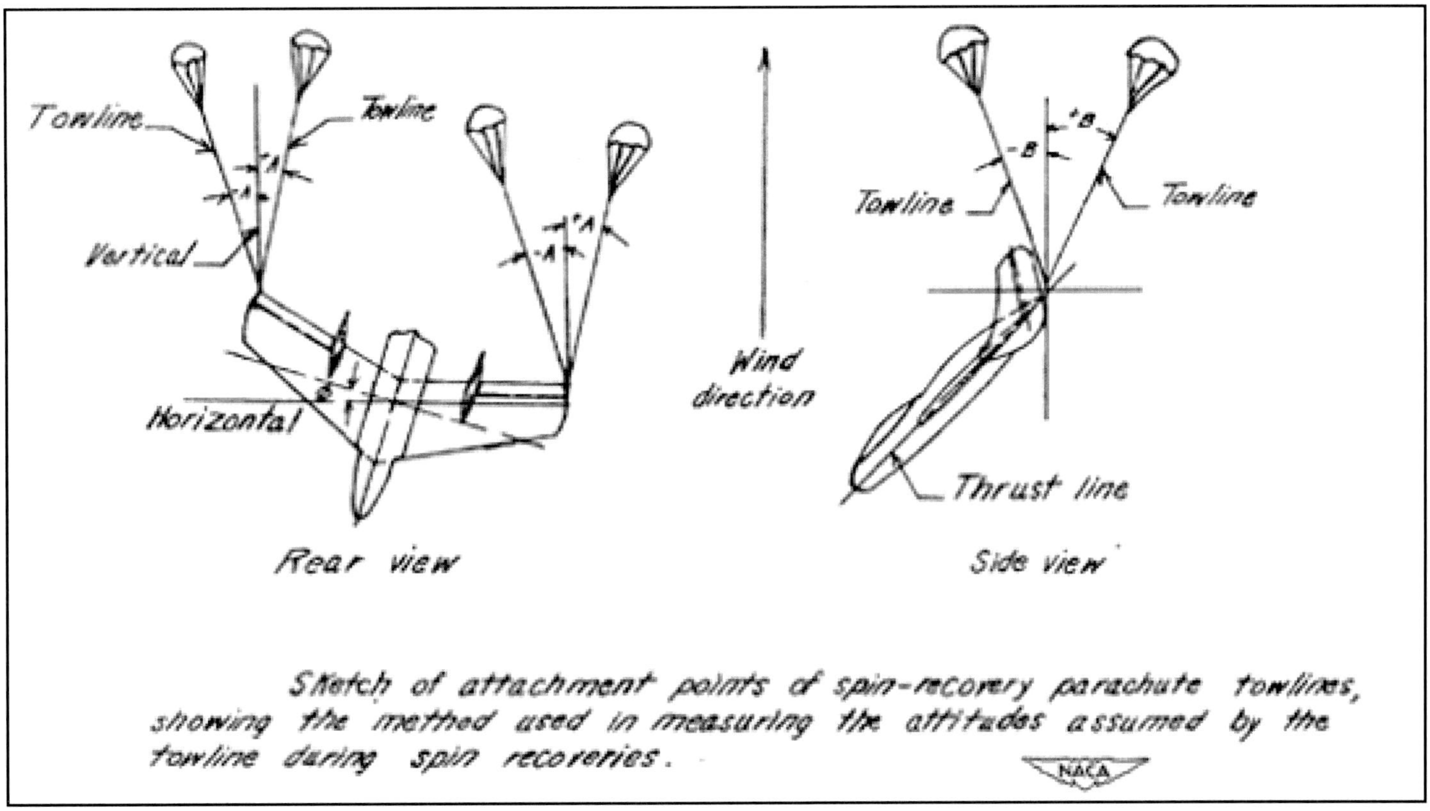

trunk lid on the aft fuselage, a canard-like surface that was deployed from the sides of the forward fuselage, and extending only the upper side of the trailing-edge speed brake. A deflector plate in the jet exhaust was also considered, possibly utilizing the eyelid mechanism of the afterburner nozzle.

The trunk lid and split speed brake proved to be ineffective with respect to generating the desired nose-up pitching moment. The canard-like device was effective but destabilizing in pitch. The split rudder was the most effective but also increased drag about 30%, causing an unacceptable decrease in rate of climb after takeoff. The deflector plate was rejected because of weight, complexity, and the difficulty of providing trim control as thrust varied or any at all at low power.

The fin flap seemed the most promising and evaluated further. Depending on the size of the flap, it provided a pitching moment equal to five to ten° of up-ailavator deflection. However, it added about 250 lbs to the empty weight, moved the c.g. aft about 1.5% MAC, and added drag during climb.

Fortunately, the engineers determined during testing that increasing the amount of the leading edge of the wing that was slatted from 60% to 100% provided the requisite lift and only required 25° of the total up-ailavator travel of 30° for trim. Adequate roll control remained. The new inboard slat had to be reduced in chord, however, in order to retain the size of the fuel cell in the forward inboard section of the wing. The original slat on the inboard wing had stopped just short of the fuel cell, extending only a short distance inboard from the fold joint.

The weight increase of the new slat configuration was estimated to be approximately 100 lbs, which reduced takeoff rate of climb by 30 feet per minute and increased stalling speed and takeoff distance by only 0.3 mph and 30 feet, respectively. However, it also increased the angle of attack required to achieve the additional lift.

Below, In an effort to obtain more pitching moment to maximize the angle of attack that could be attained on takeoff or landing, several different auxiliary pitch-trim concepts were evaluated in wind tunnel testing in late 1946. None proved to be satisfactory. (Author)

V-346 Auxiliary Pitch Trim Devices

Trunk Lid

Canard

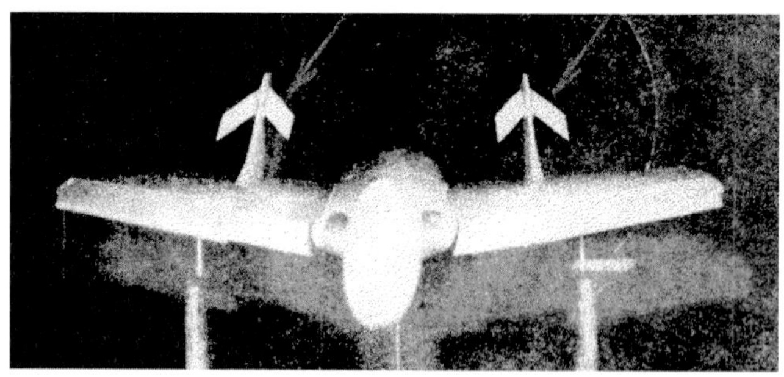

Fin Flap

Upper Dive Brake

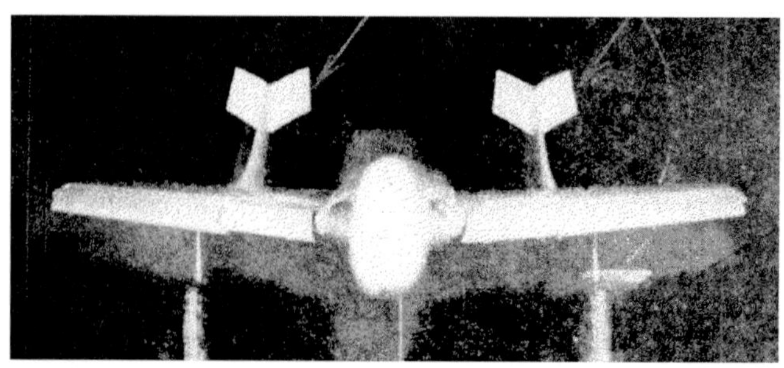

Split Rudder

Another Ton of Weight is Taken In Stride

The Navy and Vought, working together, had managed to get through the mockup review milestone without a significant gross weight gain by offsetting empty weight gains with useful load reductions. Not unusually, the weight empty growth not only continued but also accelerated as Vought further detailed the design and incorporated the results of structural analyses and wind tunnel tests. Within a year, the empty weight had grown by more than a ton, to 11,297 lbs. Most of the increase, about 1,400 lbs, was in the Vought airframe structure and landing gear. The government furnished engines and afterburners, however, had increased in weight by about 360 lbs and the surface controls by an equal amount. The weight of the hydraulic system had doubled.

At this point, weight savings opportunities were limited. About 75% of the Vought-responsible weight was estimated from released drawings and another 20% from layouts in progress. Each change was considered unavoidable, for example extending the inboard slats all the way to the fuselage, lengthening the main landing gear for greater ground clearance of the fins, strengthening the nose landing gear attach structure, etc.

Some of the weight increase was attributable to a refinement of the flight control system design. The ailavator flight control system was changed from being simply boosted to irreversible, meaning the mechanical linkage from the cockpit controls was not connected directly to the flight control surfaces. Instead, the control linkage from the pilot's stick moved a valve on the hydraulic actuator that moved the control surface. When the control surface reached the angle commanded by the pilot, a link to the actuator stopped its movement. This change was determined to be necessary because of the prospect of large and rapidly changing air loads on the control surfaces in transonic flight overpowering the boost actuators and back-driving the cockpit controls.

The incorporation of irreversible controls, however, also eliminated the feedback from the control surfaces that the pilot used to determine and modulate how much control he was applying in a given situation. An artificial feel system therefore had to be incorporated to provide that important feedback.

Pitch control feel was simulated by using ram air from a small scoop under the nose. This air was used by the pitch control feel system in con-

Below, one of the contributors to the weight growth between the mockup review and completion of detail design was a significant increase in flight control complexity when the decision was made to make the flight control boost irreversible. (U.S. Navy, National Archives)

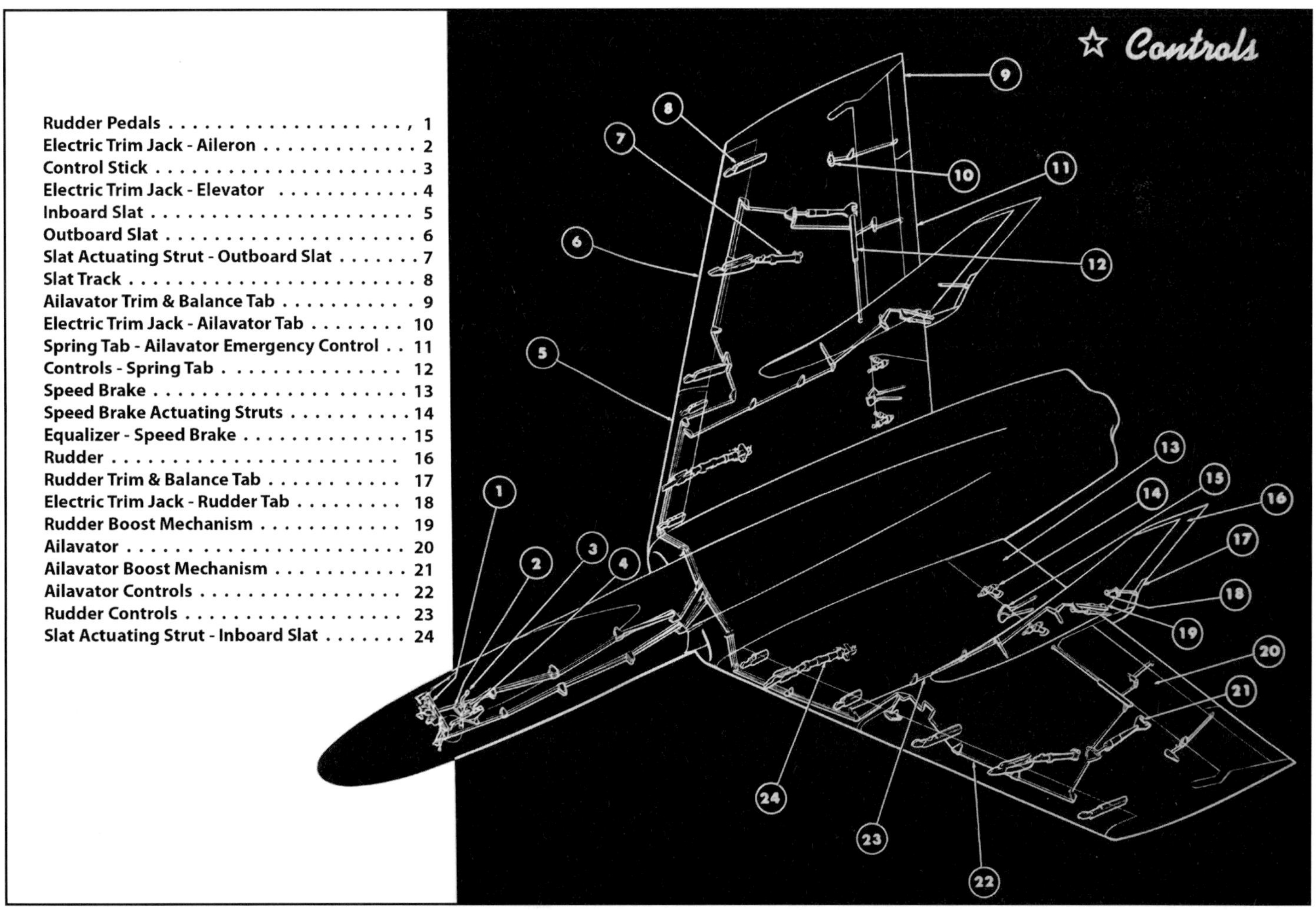

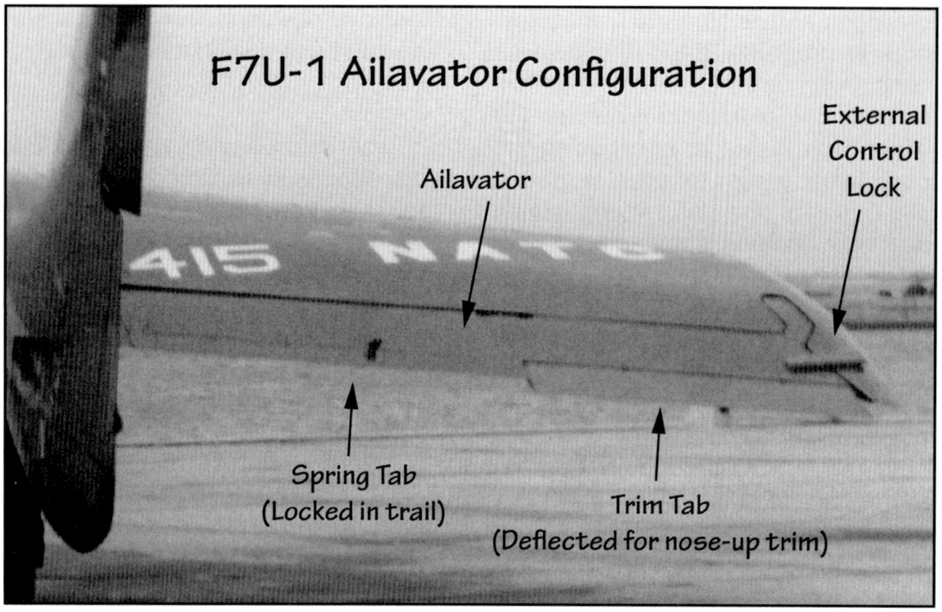

At left, there were now two sets of tabs on the ailavators, with the outboard one for trim and the inboard one for boost-off flight control. (Author)

The X7U-1 required a second rebalance due to the weight increase. It was accomplished by extending the cockpit farther forward and reducing the sweep on the vertical fins. (Author)

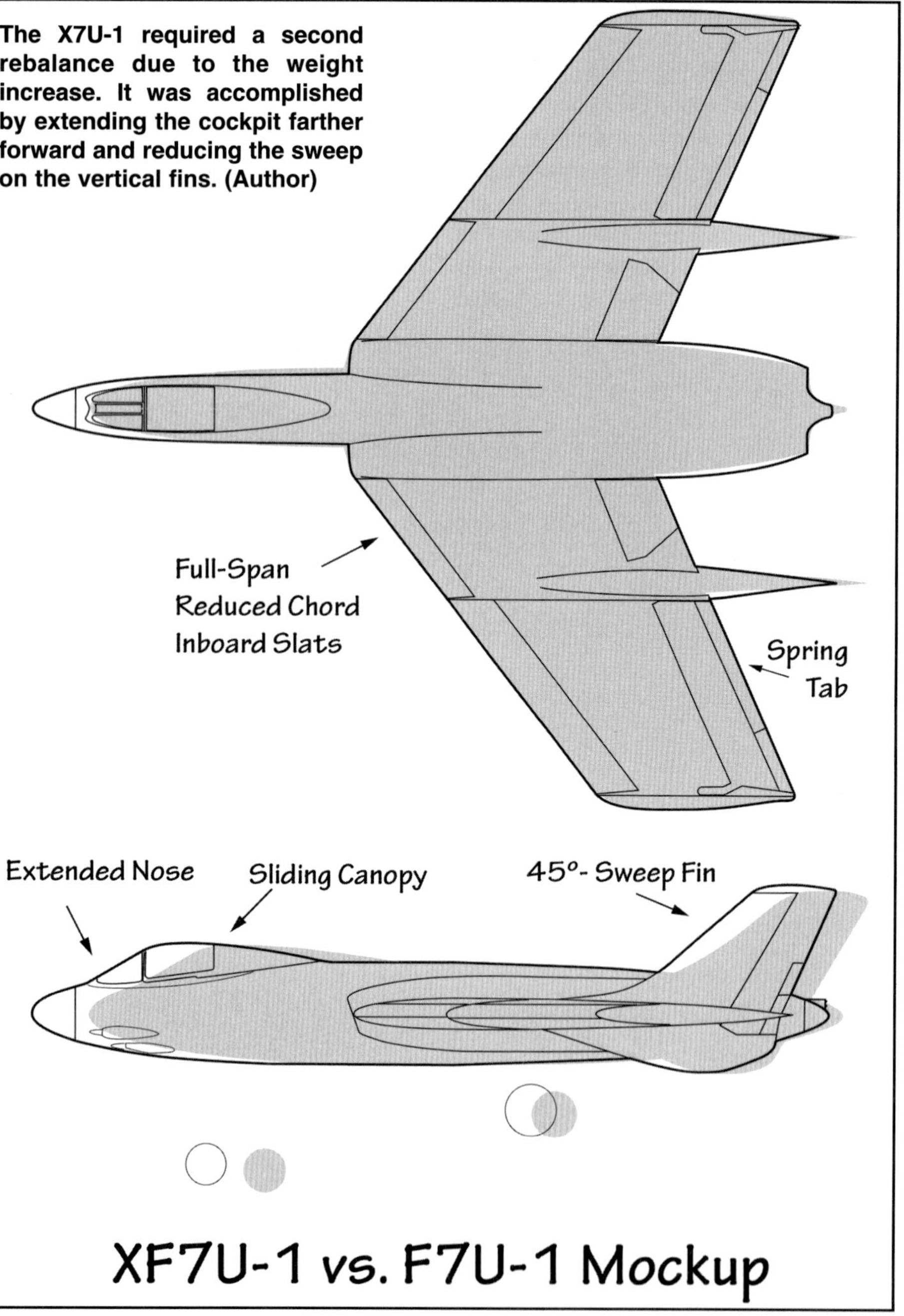

junction with stick displacement to simulate the higher stick loads that would normally result from higher speed flight and maneuvering. Roll control feel was more simply introduced with a stick centering spring that provided the same force for a given stick deflection regardless of airspeed. The rudders were originally boosted but no feel system was provided since yaw control was not believed to need it.

Further increasing the complexity of the flight control system was the requirement that the airplane still be flyable if the hydraulic system failed for any reason. In the original design, the cockpit controls were connected directly to the flight control surfaces; if hydraulic pressure were lost, the pilot lost the boost but if he was strong enough and the airplane was flying slowly enough, he could still move the controls. With irreversible actuators, when hydraulic pressure was lost the cockpit controls no longer moved the flight controls.

The simple but heavy solution to this would have been control actuators that were powered by two different and completely separate hydraulic systems. Instead, the F7U-1 flight controls were designed to automatically revert to another form of unboosted control, with the ailavators moving in response to the movement of a spring tab on the trailing edge of the control surface. When there was hydraulic pressure, the spring tabs were locked in place; when hydraulic pressure was lost, they were automatically linked directly to the pilot's flight controls and free to move. Now when the pilot moved the stick, instead of commanding the hydraulic actuators to move the control surfaces, he was positioning the spring tabs to move the control surfaces aerodynamically.

In spite of the weight growth, speed performance was still to be met, albeit in part due to a demonstrated thrust of 3,240 lbs for the Westinghouse J34-WE-22 (24C-4B) engine being provided for the F6U with a potential increase to 3,500 lbs with the Westinghouse 24C engine. The increased weight did affect the rate of climb and the stall speed estimates.

The increase in weight of engines and the control surfaces resulted in an unacceptable rearward shift of the center of gravity. Vought therefore moved the wing aft on the fuselage and extended the nose forward to relocate the center of gravity properly with respect to the center of lift. The overall length increase was partially offset by reducing the sweep of the vertical fin to 45 degrees, still acceptable from the standpoint of transonic handling qualities.

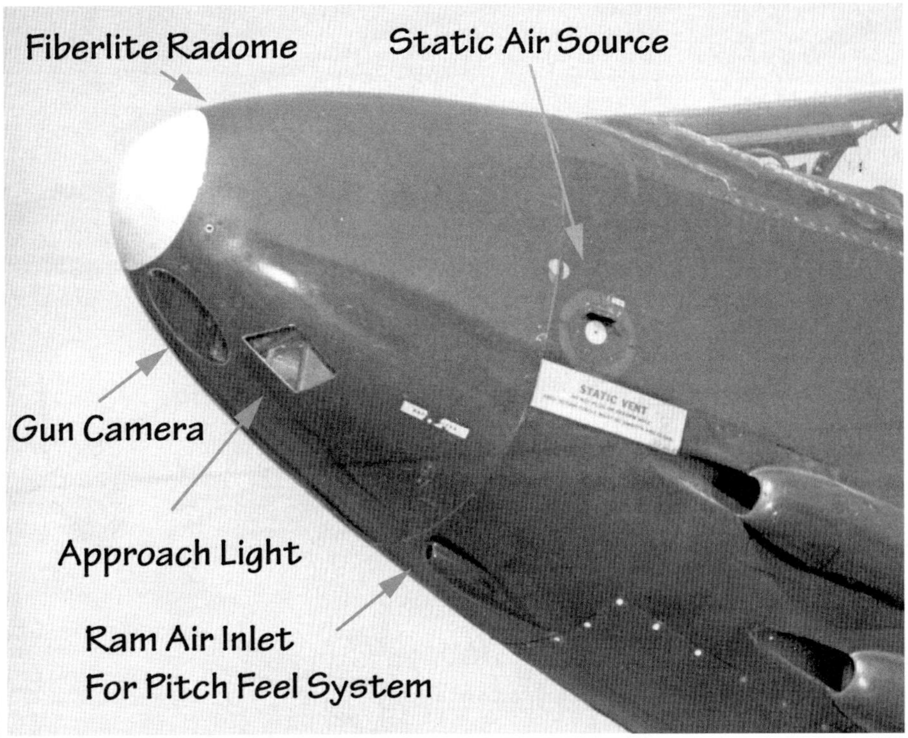

F7U-1 Nose Features

On 27 May 1948, BuAer reported that the F7U-1 was "a ton overweight, but according to Vought strong enough for carrier landing. Should be able to negotiate specification in a few days."

Although the weight increase was disappointing, the impact on BuAer's enthusiasm for the airplane was minimal because of the overall performance improvement relative to existing Navy fighters: according to CDR Clifton at BuAer, "without a doubt, the hottest thing known today."

The increased weight would normally have a direct impact on ability of the structure to meet specification requirements such as maximum g-capability and sink rate on landing, among other critical conditions. Vought suggested, however, that since most of the increased weight was in the wings, it largely offset the wing bending moments imposed by gs and that the customary design conservatism would probably account for the remainder. With respect to landing loads, Vought argued that BuAer was no longer requiring the 10 percent margin previously required by specification SS-16B. That would similarly offset the shortfall there, the rest being taken care of, again, by an engineering propensity to be conservative. Vought put off projecting any weight increase required to accommodate the arresting and catapult load increase, stating that "the contractor is considering proof loading these hooks for the higher loads in order to insure that they are satisfactory for evaluation of the airplane."

Gone to Texas

Chance Vought was founded in 1917 in Long Island City, New York. It was acquired by the United Aircraft & Transport Corporation in 1928 and moved to East Hartford, Connecticut, in 1930. It was subsequently relocated to Stratford, Connecticut, in 1939 and merged with another United Aircraft division, Sikorsky. Flight test was accomplished from the Bridgeport Municipal Airport; its runways were less than 5,000 feet long but adequate at the time.

Because of the concerns about the suitability of the runway length at Bridgeport for jet operations, the Navy and Vought agreed that the prototype XF6U would be flight tested at Edwards AFB. This change of venue paid off on the very first flight on 2 October 1946 when the engine accessory drive failed only 24 minutes after takeoff, forcing an immediate landing.

There were other difficulties with the Stratford location as well. The expansion of the plant to meet wartime production demands was ad hoc and not conducive to the most efficient operations. New England weather adversely affected flight test operations from time to time. (In 1940, the pilot of the XF4U was forced to crash land on a golf course due to rapidly deteriorating and severe weather.) As a result, the Bureau of Aeronautics suggested to Vought management that the company be relocated to a large and newly vacated government factory near Dallas, Texas. After site visits and an in-depth evaluation of the opportunity, Vought management decided to do so.

The Army established Hensley Field in Texas in August 1929 on land purchased by the city of Dallas to offload Army pilot training from Love

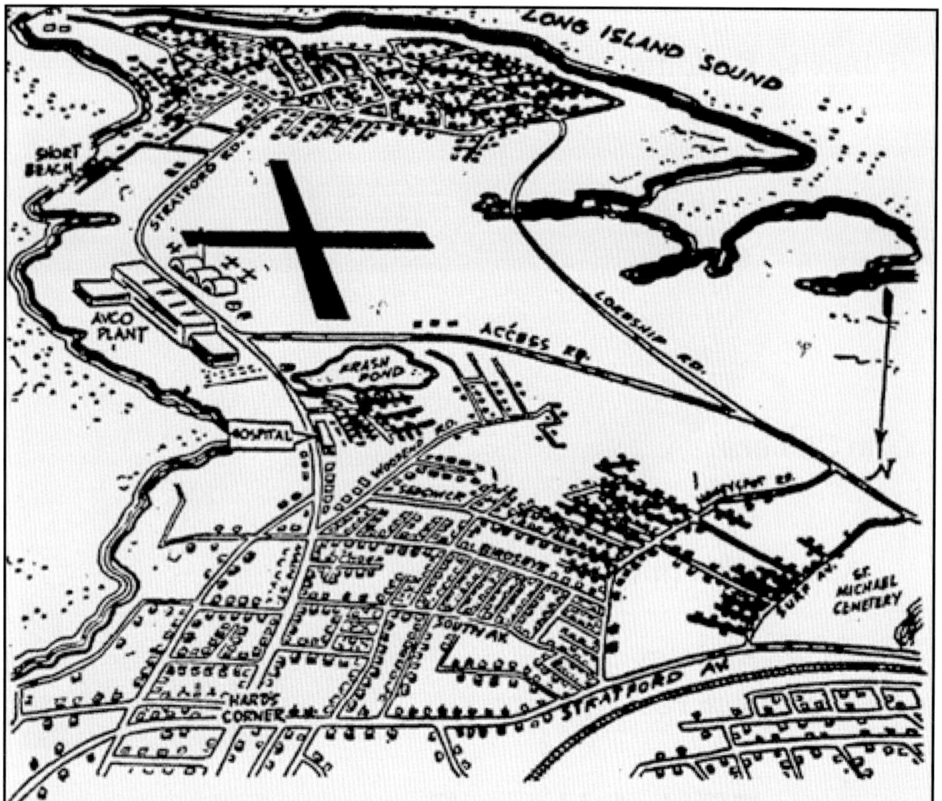

PLANE PILOT'S VIEW from glidepath approach to Bridgeport Municipal Airport in Stratford. Herald artist Ray Dirgo shows densely populated areas at North and South ends of the runway which are basis of protest by Lordship Improvement Assn., against proposal to divert air traffic from New York to Bridgeport. Stratford's Chamber of Commerce maintains that approaches and take-offs were over water and barren lands.

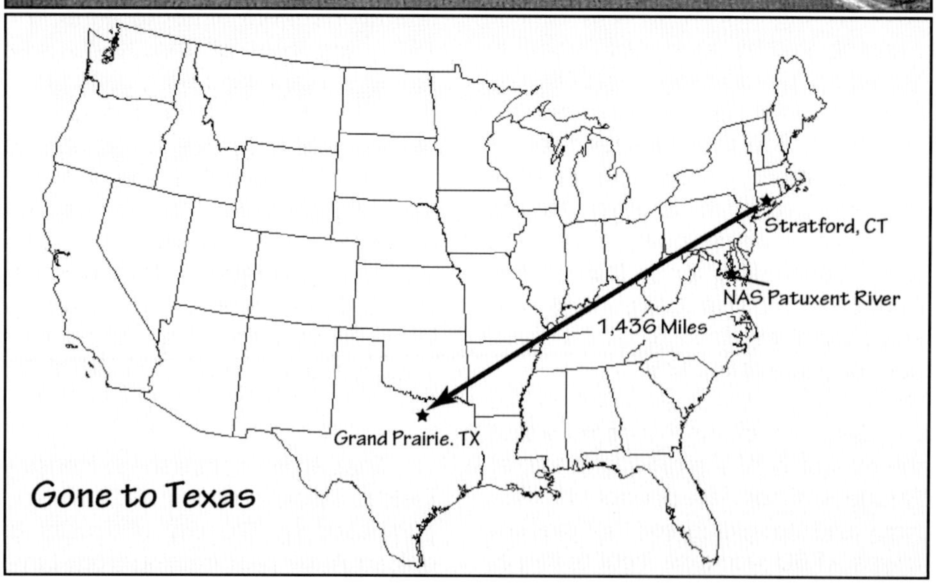

Field, the commercial airport serving Dallas. It was named for Maj. William N. Hensley, who was on the first trans-Atlantic dirigible crossing in 1919 and an Army flying instructor who had helped select the site, located southwest of downtown Dallas and adjacent to the town of Grand Prairie. North American Aviation began to build an aircraft manufacturing plant adjacent to the Field in 1940. However, responsibility for the plant was soon taken over by the Defense Plant Corporation (DPC), a government entity, which completed it and leased it back to North American. A Navy training installation adjacent to Hensley Field became Naval Air Station, Dallas, on January 1, 1943.

North American ceased production at the plant in August 1945 as a result of the cancellation of airplane production contracts following the end of World War II. Responsibility for the manufacturing facility was formally transferred to the Navy in 1948; on 18 April, Vought management publically announced its move to Texas. (The Air Force transferred responsibility for Hensley Field itself to the Navy on September 30, 1949.)

The first three production F6Us were to be built in Connecticut but taken to Texas for first flight; the remainder were to be assembled in

---

Top-to-bottom: One of the reasons that Sikorsky elected to depart from Connecticut was the fractious nature of the citizens in the vicinity of the Bridgeport Municipal Airport with respect to operations and improvements there. (Lordship History Website) NAS Dallas circa 1947, looking west northwest. The large manufacturing plant is located on the west side of the runway complex. The control tower and main ramp were located on the southeast corner. The north-south runway was to be lengthened to 7,500 feet by extending it south on landfill in Mountain Creek Lake. (Naval Aviation News, September 1947) The Vought move from Stratford, CT to Grand Prairie, TX took approximately a year with unavoidable disruption of F6U production and F7U development. (Author)

Texas from parts that had been produced in Connecticut. The three prototype XF7U-1s were to be completed at Stratford and the first one was to be flown at Muroc in California as the XF6U had been. Fabrication and assembly of the production F7Us was to be entirely accomplished in Texas.

Almost all of the machinery and equipment in the plant, engineering laboratories, and offices; about half the engineers; and many of the rest of the employees were to be moved between July 1948 and June 1949. It was reportedly the largest industrial relocation of its kind up to that time, involving 27 million pounds of equipment and 1,300 employees and their families.

The first production F6U was trucked to Dallas in February 1949; it flew for the first time in July 1949, surprisingly well ahead of schedule particularly given the changes made from the prototype like the incorporation of afterburning, not to mention the disruption caused by Vought's move from Connecticut to Texas.

The runway at NAS Dallas was only 5,200 feet long, inadequate for testing jet aircraft. It was therefore being lengthened to 7,500 feet. Although F7U flights to and from NAS Dallas were accomplished from time to time before the runway extension was completed, flight test would first be accomplished out of Carswell AFB. Vought subsequently established a flight test operation at Ardmore, Oklahoma, a former World War II training airfield about 100 miles north of Dallas. (The ground-loving F6Us were trucked to Ardmore for their first flights and deliveries.) Vought was finally able to relocate the F7U flight test operation to Dallas from Ardmore in mid-May 1950.

Above, while the main runway was being lengthened at NAS Dallas, F7U-1 flight test was initially accomplished at Carswell AFB, which was located west of Fort Worth and about 28 miles from the Vought plant. (Don Pyeatt collection) Below, the weather in Texas was virtually ideal for flight test. Moreover, the Dallas-Fort Worth area was far less built up than Connecticut in general and the Stratford area in particular, reducing the likelihood of noise complaints and the risk of collateral damage if an airplane had to be abandoned in flight. Here, the XF7U is landing at Carswell AFB with Air Force B-36s in evidence. (Don Pyeatt collection)

XF7U-1 Initial Flight Test Patuxent River

The Navy originally intended that the XF7U first flight would be accomplished by Vought at "Muroc Dry Lake", the future Edwards Air Force Base, like the F6U's had been. In early 1948, however, BuAer decided to have Vought make the first flight at Patuxent River to save time and money. That it was a Naval air station rather than an Air Force base may have been a factor.

The first XF7U was completed on schedule in Stratford, Connecticut, in July 1948 and transferred to flight test for systems checks. Although first flight had been expected by mid-August, an error was discovered in the design of the fin attachment to the wing, requiring its modification. After checkout, concluding with taxi tests at Stratford beginning on 15 September, it was transported on a self-propelled, ocean-going, barge-type vessel down to NAS Patuxent River, Maryland.

Although Solar Aircraft Corporation had already delivered at least one afterburner to Vought for installa-

Above, on 9 September 1948, the first XF7U-1 was rolled out, essentially complete but only partially painted, for initial taxi tests at the Bridgeport Municipal Airport. (Vought Heritage Center) Below, for some reason, the vertical fins were not identical in basic construction. The Fabrilite section of the left fin extended farther down than the one on the right, probably because the 3-in-1 antenna was located inside the left fin. However, that suggests that the Fabrilite structure was heavier than a metal counterpart. (Vought Heritage Center) Bottom, an article on the XF7U in the 29 November 1948 edition of Aviation Week likened the landing gear to that of a DC-3: "Unusually heavy construction embodied in the tricycle landing gear gives a clue to the relatively high gross weight of the Cutlass and the tremendous impact the gear must take on carrier landings. Wheels are of particularly sturdy construction. Shock struts are more nearly the size of those on a DC-3 than those normally associated with fighter type construction." The nose gear appeared to be particularly robust. (Vought Heritage Center)

tion in a modified XF6U, none were available for the first flight of the XF7U because BuAer had not made a timely decision as to whether the F7U would have Solar or Ryan afterburners. In simulated altitude tests at NACA Cleveland, the Ryan afterburner was superior in efficiency; the Solar test ended with a flame-holder failure after 12 hours and 15 minutes of operation. The question would subsequently be resolved in Solar's favor but an order wasn't placed for XF7U afterburners until August 1948 so they wouldn't be available for installation in the second XF7U-1 to fly either.

Preparation at Pax for the first flight of the Navy's new, high-performance jet fighter included installation of an emergency arresting capability that consisted of a cable attached to chains. The concern was that there might be a boost control failure in flight requiring a high-speed landing. It became even more important because the leading edge slats were not yet functional.

The XF7U's first flight was accomplished on 27 September 1948 by Vought test pilot J. R. (Bob) Baker. For some reason, it only lasted six minutes. The second flight on 29 September 1948, also by Baker who did most of the initial flight testing, is documented by Vought as the official one. It was only a year after the first flight of the Air Force's F-86.

The Navy's new fighter was publically displayed to the press on 18 November 1948 at Patuxent River by a very senior Navy test pilot, Captain Frederick M. Trapnell, the Test Center Coordinator. At that point, the first XF7U had only been flown 26 times for a total of about 16 hours. It had not yet been flown with either the leading edge slats extended or the afterburners installed. The second Cutlass was in final assembly at Stratford.

According to the 29 November 1948 issue of Aviation Week, "Capt. Trapnell held the Cutlass down to 500 mph in level flight but gave ample evidence of its high rate of climb and relately high maneuverability."

However, at a conference with BuAer on 22 November, Trapnell made the following critical comments:

I. There did not appear to be enough up elevator to stall the airplane at sea level, as almost 90% of maximum up elevator was required to stall at altitude with the landing gear extended.

II. Elevator forces "reversed" with the control boost off.

III. Directional stability was low. Moderate aileron was required to hold zero roll in yaw.

IV. The rate of roll was not as fast as the FJ-1 or XF9F but not unacceptable.

V. The dive brakes did not seem as effective as on other airplanes.

VI. The transonic drag rise required a dive of 25-30° to reach Mach 0.92; this seemed excessive compared to other jet fighters but Vought was questioning the calibration of the Mach meter at high speed.

VII. A tendency for over control in pitch, i.e. porpoising, was occasionally evident.

He did note that the cockpit was quiet with the canopy open.

Not withstanding the overweight effect on rate of climb and ceiling, the speed guarantee appeared to be achievable. Several level-flight speed runs had been made at up to "practically full military power" (the Solar afterburners were not yet installed) and the achieved speeds were faster than predicted on three out of five flights. "(T)he thrust of the engines has not been determined by ground runs, so it is difficult to evaluate accurately the performance. Also the engine configuration has not been consistent from flight to flight so that it is possible that the speeds from flight to flight are not comparable."

The highest indicated airspeed that had been reached was 512 knots at 4,000 feet, approximately Mach .82. Higher Mach numbers had been achieved at 30,000 feet without any notable compressibility problems like

Above, following the taxi tests at Vought, the first XF7U-1 was cocooned and on 17 September 1948 loaded on a self-propelled barge for transport to NAS Patuxent River for flight test. (NMNA) Below, Vought test pilot J.R. (Bob) Baker took the XF7U-1 up for the first time very shortly after its arrival at Patuxent River. Two different dates are cited for the event. The official first flight was 29 September but a very short one reportedly occurred on 27 September. (NMNA) Bottom, BuNo 122472 landing gear extension on 18 October 1948. (National Archives)

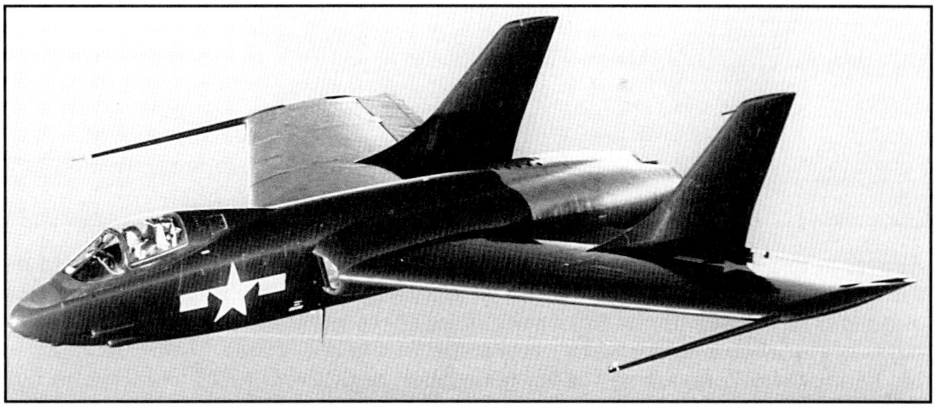

Above, a formal photographic session was accomplished on 13 October 1948. Note that the edges of the landing gear doors have been painted white for an evaluation of their complete and sustained closure in flight. (NA 80-G-396985) At left, the clamshell doors over the opening for the stinger tailhook were eliminated by having the aft side of the hook point shaped to the curvature of the boat tail. The multiple ovals on the fuselage spine are for engine compartment pressure control. The small dark areas at the forward juncture of the engine nacelles and the forward fuselage are for the discharge of part of the boundary layer air taken off at the intakes. (Vought Heritage Center)

Above, in this view of BuNo 122472, two small white stripes can be discerned on the inboard side of the left-hand fin. The lower one on the stub marked a foot step and the one half way up the fin, a hand-hold. These were for the start of the boarding process when a ladder wasn't available. What looks like rear vision windows in the canopy were provisions for the installation of motion picture cameras to film flow-visualization tufts on the wing and/or vertical fin/rudder. (NMNA) Below, Capt. Frederick Trapnell took off on 18 November 1948 for the first public demonstration of the Navy's newest jet fighter. Note that the leading edge slats have not yet been made functional. (NMNA)

airframe or control buffet. The handling qualities were considered to be good even in rough air. Before first flight, boost-off control loads had been a concern, particularly for landing, but proved to be acceptable at speeds up to 350 knots and boost-off landings were not difficult. It was recognized, however, that boost-off flight at high Mach numbers might prove to be otherwise. Although no flights had been made with the leading-edge slats extended, stall characteristics were very good and expected to be better when they were made functional. The predicted maximum lift coefficient (with slats retracted) was being achieved.

Vought chief pilot W.H.B. Millar and test pilot W.P. Thayer also checked out in 472 while it was at Patuxent River. The flight test program was off to a good start.

On 3 December 1948, Bob Baker ferried the first XF7U-1 from Patuxent River 70 nm south to NACA Langley, located on Langley AFB near Hampton, Virginia. The purpose was to calibrate the static pressure measurement at high altitudes and Mach numbers using a radar theodolite technique developed by NACA engineers. The airplane's static pressure source provided data to the altimeter, rate of climb indicator, and airspeed indicator. The total pressure measurement also used for airspeed was considered to be fairly accurate but an accurate static pressure—necessary for determination of airspeed and altitude—was harder to obtain, particularly at transonic speeds.

By this point, Vought had added an instrumentation boom to the nose in addition to the original booms at each wingtip. The wingtip booms were reportedly removed because they shook too much, but since was the 37th flight and testing had been accomplished from stalls to high EAS and Mach numbers, it couldn't have been too significant of an issue. Although resonance of the wingtip booms with airframe or aerodynamic forcing functions might have been a problem, it's likely that the nose boom, since it extended the Kollsman

Above, Vought test pilot Bob Baker flew the first F7U-1 to NACA Langley in early December 1948 for static pressure instrumentation calibration. (NASA L-58850) Below, for the pressure instrument calibration flights, the bottom of the wings and stub fins were temporarily painted white to aid in tracking the airplane visually from the ground. (NASA L-58845) Bottom, in the February 1949 issue of Naval Aviation News, the XF7U-1 was likened to a praying mantis. (Naval Aviation News)

high-speed pitot/static head farther ahead of the aircraft, provided better accuracy, particularly of the static pressure and at high Mach numbers.

NACA Langley had developed a static pressure calibration that utilized radar-phototheodolite tracking equipment and was considered accurate even if the airplane was maneuvering or diving. Airspeed calibration had traditionally been accomplished at relative low altitudes but this precluded measurements at high Mach numbers.

An SCR-584 radar combined with a phototheodolite accurately established the aircraft's altitude by combining the slant range distance provided by the radar with the elevation measured by the phototheodolite. Cameras recorded the radar-scope image, the target, and the elevation. Both the aircraft and the ground station were equipped with very accurate chronographs so airspeed/altitude measured by the aircraft instrumentation could be precisely compared to that measured by the ground station.

The technique first required a survey of the static pressure and free-stream temperature of the air in the range of altitudes to be flown but at lower airspeeds, with the static pressure system having been calibrated at low altitudes. The airplane was then flown through the altitude range at much higher airspeeds—in the case of the XF7U at up to Mach .94—to determine the difference between the static pressure measured at low speeds versus the high speeds and thereby provide a correction for altimeter and airspeed measurement.

NACA provided an extrapolation of the correction to Mach 1.0 and subsequently revised it based on the wing-flow testing.

People who *see the Navy's new jet fighter, the Chance Vought XF7U-1 Cutlass, for the first time, usually remark that it looks like a giant bug squatting on the runway. Its nose is 15 feet in the air and its tail a couple of feet off the ground.* NAVAL AVIATION NEWS *was struck by its resemblance to a Praying Mantis and herewith presents photographs of the two to show the similarity. The Mantis is about five inches long. The plane's enemies may pray, too.*

The Second and Third XF7U-1s Begin Testing

The second XF7U-1, BuNo 122473, was used for non-destructive airframe proof-load tests beginning in mid-1948 so it would be the third, not the second, XF7U-1 to fly. Instead, the third XF7U-1 built, BuNo 122474, was the second to be flown, on 13 December 1948. Now that Vought had some F7U flight experience, 474 began its first flight at Stratford, Connecticut, but landed 230 nm southwest at NAS Patuxent River, Maryland.

By the end of the 1948, BuNo 122472 had been flown more than 50 times in four months. BuNo 122474 was also off to a good start, having flown eight times in only two weeks. Nevertheless, flight test was somewhat behind schedule. The Navy attributed most of the delay to government-furnished engine governors and the lack of afterburners. Part 1 of SR-38D, Structural and Aerodynamic Demonstration, was roughly one-third completed and according to a BuAer status report, "the demonstration program is being vigorously pursued. No serious aerodynamic deficiencies have been brought to light in this program."

In January 1949, once initial flight test and evaluation had been accomplished at NAS Patuxent River, BuNos 122472 and 122474 were ferried together to Vought's new home in Texas. Fuel consumption tests were first accomplished at Pax, including a test and a subsequent modification of the fuel transfer system as well as fuel consumption checks at 20,000 and 30,000 feet. The final evaluation was a simulated ferry flight lasting two hours and 13 minutes.

Because of weather and a lack of knowledge of winds aloft at cruise altitude, the actual ferry flights were only about an hour in duration. The first leg was from Pax to Shaw AFB, South Carolina, on 8 January 1949. The second was from Shaw to Maxwell AFB, Alabama, the same day. The third was from Maxwell to Barksdale AFB, Louisiana, on 10 January. The last leg of the ferry flight was delayed by weather until 19 January. The first three legs averaged about 400 statue miles each and the last was less than 200 miles, so there was probably an impromptu air show on arrival at Vought's new facility in Texas.

Above, the second XF7U-1 to fly, BuNo 122474, joined the first, BuNo 122472, at Patuxent River in late December 1948. In this 4 January 1949 picture, distinctive markings are now evident on their noses. (This was to become standard for F7U-1s on Vought test status.) In January, they were ferried together to Vought's new home in Texas. (NA 80-G-397968) Below, BuNo 122474 shortly after completion. Note small tuft camera window aft of the canopy and the lack of case/clip chutes. (Chance Vought)

XF7U-1 472 landed at NAS Dallas to go into layup for installation of afterburners and functional wing slats. XF7U-1 474 continued a few more miles westward to Carswell AFB where Vought would initially conduct Cutlass flight tests in Texas.

Initial Development Testing

One of the first tests at Carswell with 474 was evaluation of carrier-type landings. At the time, they were very similar to short-field landings.

The pilot made his approach to the runway flying just above the stall speed, maintaining a low and constant altitude with power and following the guidance of a Landing Signal Officer (LSO) as to line up, height, and speed. The LSO would give the pilot a signal to "cut" the engine at the point when a power-off descent would result in a touchdown past, but not far past, the aft end of the simulated carrier deck.

Vought pilots made a total of 62 simulated carrier landings, all with the slats closed, in this initial evaluation. Yellow markers were placed on the runway where the arresting wires would be. Cameras recorded the approach and landing. Vought test pilots Bill Millar, Bob Baker, and Paul Thayer took turns making the carrier approaches at a wheel height above the ground of about 20 feet while a Vought engineering trainee, Bill Sunday, who had been a Landing Signal Officer (LSO) on *Essex* when he was on active duty in the Marines, did the waving. (He later became a Vought test pilot and was killed flying an F8U Crusader.) In no case was the fin stub clearance with the runway at touchdown less than 12 inches, which was reassuring.

Most of the requirements of SR-38D Part I for the gear-up conditions were also accomplished with 474 in February. Since the pilots thought that the rudder sensitivity was too high, the rudder boost actuators were removed and replaced by a mechanical linkage. This not only reduced rudder sensitivity to a more appropriate level, elimination of the irreversible actuator provided stall warning from rudder buffet transmitted to the rudder pedals approximately six to seven knots before the stall. A problem that proved more difficult to solve was longitudinal stability with an aft cg: as Captain Trapnell had noted, it resulted in a tendency for the airplane to porpoise and the pilot to over control.

474 was flown from Carswell to the Vought plant at NAS Dallas on 24 March 1949 for its slat activation and afterburner installation, swapping places with 472. Among 472's first assignments after arrival at Carswell on 29 March was the completion of the Part I Demonstration requirements and flights at high Mach numbers using the afterburner.

Meanwhile, BuNo 122473 had completed the proof load testing at Stratford required to clear the airplanes structurally for flight test at design limit loads. It was first flown at Stratford on 15 February 1949 and ferried to Patuxent the same day. The first major XF7U-1 incident occurred on 18 February 1949 when Vought chief experimental test pilot William H. B. Millar had to make a gear-up landing at Patuxent in 473 due to a failure of the wheels-up selector valve. However, the main gear wheels had extended about six inches before they jammed and the brakes were functional, so the damage was relatively minor, limited to the landing gear doors and fin stubs.

The damage was quickly repaired and the airplane resumed flight tests on 5 March for a field carrier-landing evaluation with the wing

Above, some of the first flight tests accomplished in Texas involved simulated carrier landings as depicted in this clip from an old Vought internal publication for employees. (Vought Heritage Center) Below, fresh from proof testing and now the paint shop, BuNo 122473 has been rolled out, albeit without its engines, accounting for the unusual stance.

slats extended. Clearance of the fin stubs was less than with the slats retracted since the slats allowed flight at a higher angle of attack but still satisfactory. A total of 17 landings were made before tragedy struck.

Millar was killed on 14 March 1949 while flying 473 on its 13th flight. He had accomplished a post-maintenance shakedown flight and then a planned air-to-air photography session flying formation with a Beech SNB and also flying by it. That requirement completed, he broke away from the slow propeller-driven transport to proceed independently (according to Vought's accident summary, "presumably entering (the) pattern to land") and was never seen again. From what little wreckage was subsequently recovered from the Chesapeake Bay after an extensive search, Vought engineers determined that the Cutlass had crashed into the water with the slats and landing gear extended in a left roll but were unable to establish a cause. It turned out to be a harbinger of a fundamental control system design shortcoming.

The modifications to 474 were completed in early May 1949. It was ferried to Patuxent on 13 May to

Top-to-bottom: This photo might have been taken on 473's last flight before it and Vought chief experimental test pilot William H.B. Millar went missing almost within sight of NAS Patuxent River. Note the sticks extending from the bottom of the fin stub. After landing, these indicated how much clearance there was between the fin and the deck on touchdown. (Vought Heritage Center) The third XF7U-1 to fly, BuNo 122473 was not on flight status long before a hydraulic component failure required a partial gear-up landing. The damage was basically limited to the landing gear doors and was repaired within two weeks. (Vought Heritage Center) A photo taken of 473 after it had been lifted up off the runway following its partial gear up landing shows the fairing attached to the hook point that conforms to the shape of the aft end of the fuselage. (Vought Heritage Center)

replace 473 in the carrier suitability test program of field-arrested landings and catapult launches. It was then ferried west again, this time to Ardmore, Oklahoma,

A Revisit of the Carrier Takeoff and Landing Design Requirements

In August 1948, a Naval Reserve Officer temporarily assigned to the Fighter Class Desk calculated that the F7U-1 was at risk of hitting its tail on landing. (VF-5A had unexpectedly experienced tail bumping during FJ-1 carrier operations earlier in the year and the Douglas F3D was also bumping its tail.) He concluded that a flatter approach to minimize the incidence of lower-fin strikes would result in too high a landing speed. He also noted that as high as the nose was now, an even higher pitch attitude would be desirable to minimize takeoff distance.

It turned out that Vought was aware of the issue, among others, and proposed changes were in work for that and other issues associated with operation from an aircraft carrier.

The landing aboard the carrier was of particular concern. As long as the nose gear was, there was still a possibility of a significant pitch-down caused by a combination of excessive nose-up attitude, a high sink rate relative to the deck, and the arresting hook picking up a cable while the airplane was still in the air. This could result in the nose wheel hitting the deck harder than the structure to which the nose landing gear could accommodate. In order to reduce the pitching moment generated by the

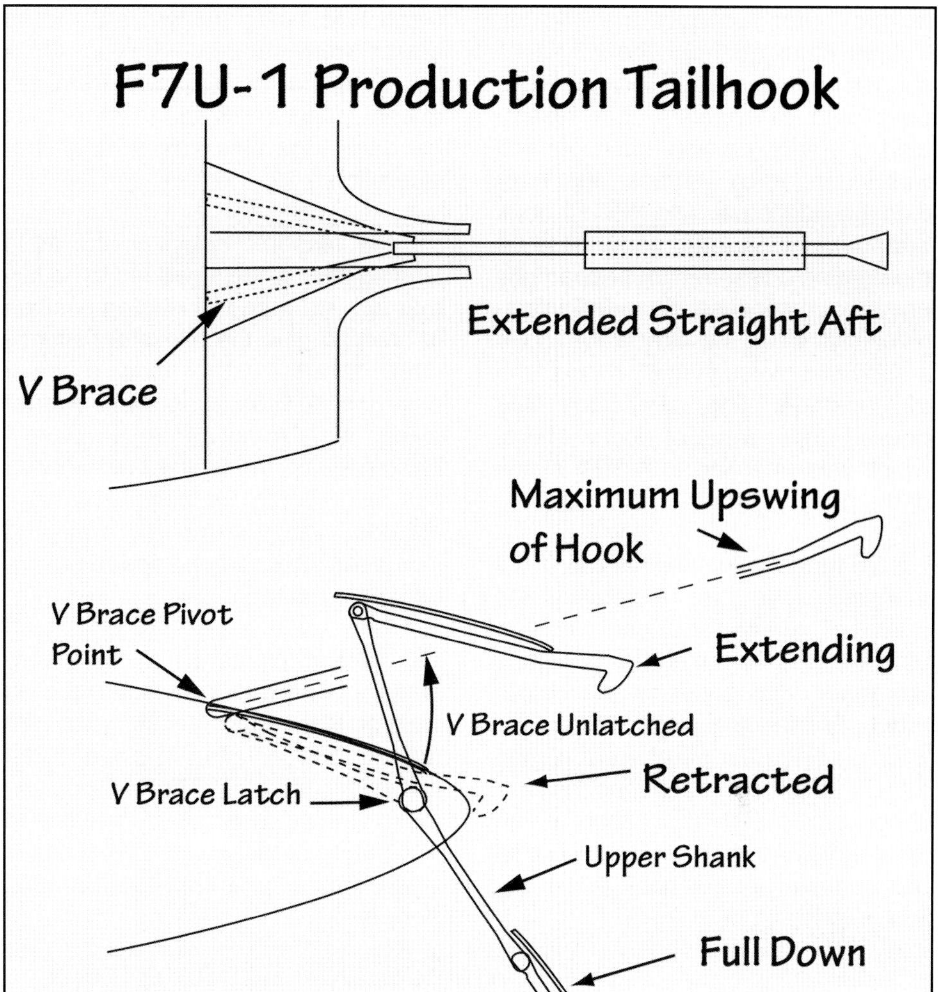

In order to raise the attach point of the tailhook as high as possible to minimize pitch-down loads on pendant engagement, Vought engineering created a triple-jointed contraption that folded into a small space at the top of the aft fuselage. (Author)

At right, the production hook retained the cable shedding feature of the XF7U-1's stinger hook. The tailhook installation change resulted in an increase in the size of the fairing between the engine exhausts and the relocation of the orange and white rearward-facing position lights, which had been a single unit located immediately above the stinger tailhook opening, to a fixture on each side of the aft fuselage. (Vought Heritage Center)

tailhook, for production F7U-1s, Vought changed from the XF7U-1's stinger hook to a complicated, double-jointed mechanism that resembled a trench-digging backhoe and was extended by V-braces that handled the side loads. The XF7U-1 hook arrangement provided some nose-up moment on arrestment but apparently not enough because the redesign significantly raised the attachment point of the hook by locating it on top of the aft fuselage. The new hook was housed under a set of doors, one of which was attached to the aft shank of the hook.

To further minimize the risk of tail bumping while also increasing the angle of attack and pitch control effectiveness for takeoff, a two-position main landing gear was incorporated. For landing and parking, the main landing gear struts were angled aft. On landing, this reduced the likelihood of a tail strike in a nose-high touchdown. As importantly, it reduced the potential for tip back if the airplane was being pushed aft and its brakes were suddenly applied. On a carrier, since the airplanes were frequently parked with the tails out past the deck edge, this was a major consideration.

For takeoffs, however, the main landing gear wheels needed to be farther forward than the landing position in order to give the ailavators slightly more leverage, providing a lower nose wheel liftoff speed and shorter takeoff run. From a carrier compatibility standpoint, the forward wheel position also slightly increased the nose attitude at launch, providing increased lift at the end speed provided by the catapult.

Several flights using 474 were made at Carswell to evaluate the aft position of the main landing gear, with a fixed strut installed in place of the retraction actuator. The two-position main landing gear was installed in 474 during a mid-July to mid-October layup that also added afterburners, new speed brakes, and control system changes.

The nose landing gear was also redesigned for production. First, concerns had arisen about its length causing high loads in the fuselage structure on landing. When the nose wheel touched down, the drag load resulting from friction between the tire and the deck was magnified by the length of the strut. In order to reduce that initial rearward (kick-back) load, a pre-rotation device was added to the nose wheel. It consisted of a waterwheel-like device on the nose

Below, Vought introduced a two-position main landing gear to reduce the risk of tail bumping on landing while also retaining a reasonable nose-wheel lift-off speed. The nose-wheel prerotation feature reduced the kick-back load on the nose landing gear actuator and mounting structure. (Author)

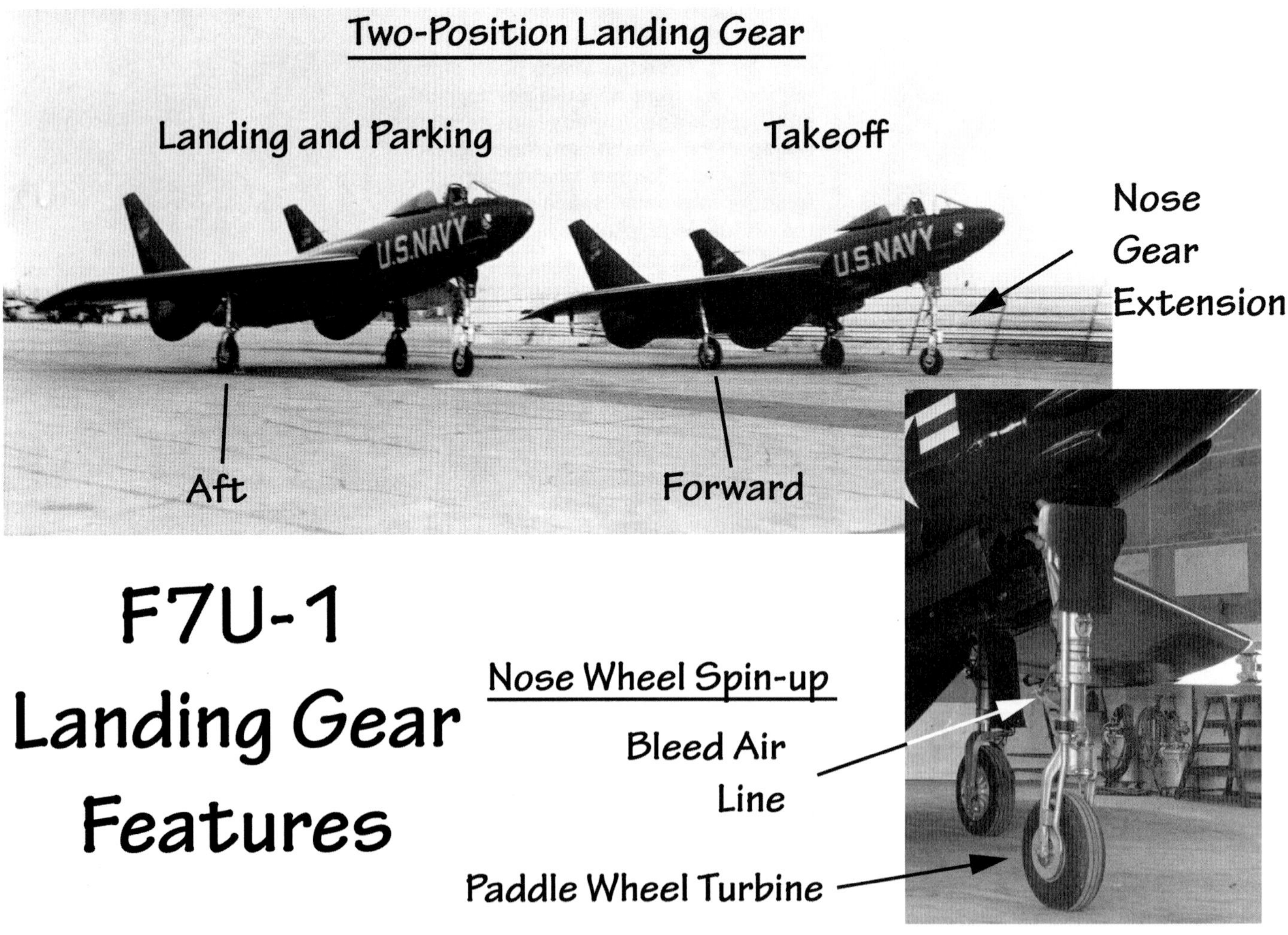

wheel hub that was rotated by engine bleed air piped down to the wheel. The wheel rotation meant that the initial rearward loading was caused by the tire's rotating friction, not the higher sliding friction. The strength of the nose gear itself was increased by changing the fork holding the nose wheel from single-sided to two-sided.

As the weight increased, the need for a higher angle of attack on takeoff, which Vought had originally thought was not necessary, became more apparent. One of the solutions, sketched out in March 1949, was the pedestal launch concept. This put the nose wheel up on a raised platform as part of hooking up the airplane up to the catapult shuttle and holdback. A more practical approach, albeit

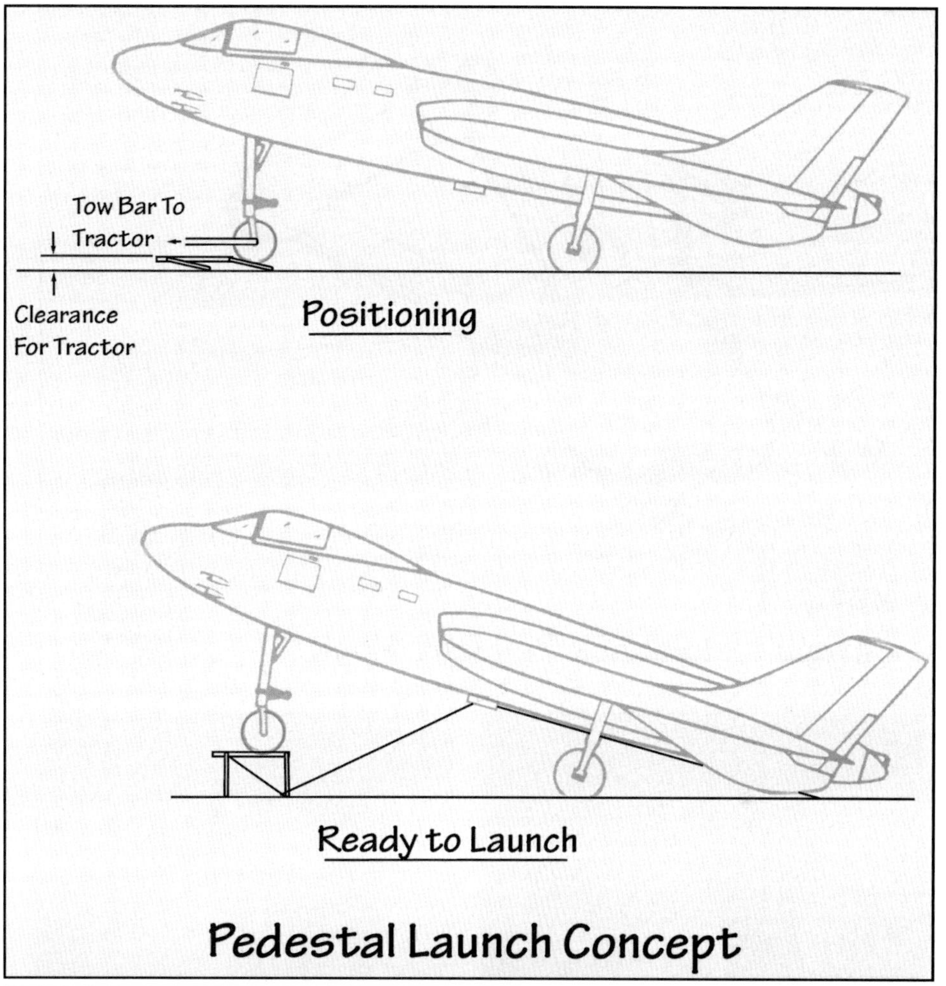

Above right, Vought sketched out this pedestal launch concept in March 1949 as one means of positioning the F7U at the angle of attack for maximum lift for launch. Note that a tow tractor wouldn't be required to move the airplane into position; before nose gear steering was incorporated, a member of the catapult crew would help maneuver the airplane precisely into position for launch with a steering bar hooked into the nose wheel. (Author) At right and below, F7U main gear details. (Tommy Thomason collection)

heavier, was to extend the nose gear for launch. The strut extended 15" when the retractable catapult hook was pulled out by the ground crew. The pressure that caused the extension was automatically relieved when the gear retracted.

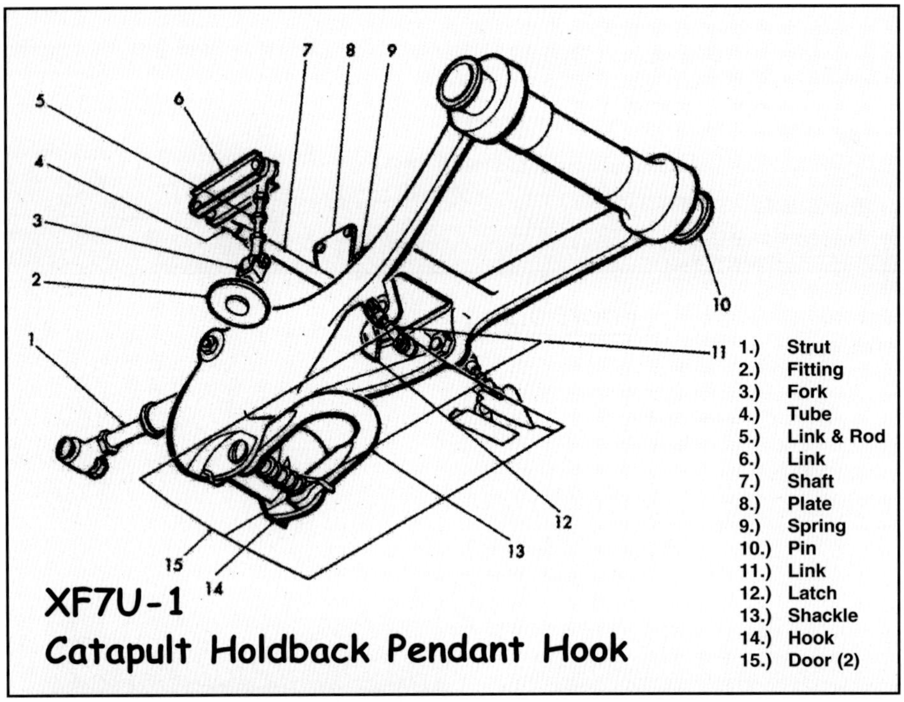

At left, the catapult holdback hook was located in the center of the fuselage at station 210.8 behind two small catapult hook doors. (Chance Vought) Below left, further analysis of the nose landing gear loads led to the incorporation of the prerotation feature and a two-sided fork. (Author) Below, right-hand view of the production F7U-1 nose gear which shows the pipe-like nose wheel pre-rotation air line and double-sided fork to good advantage. (Tommy Thomason collection)

Prototype

Production

F7U-1 Nose Landing Gears

OS-113 Interceptor Competition

Even before the XF7U-1 had flown, Vought proposed a derivative of it, the V-362, for BuAer's Outline Specification 113. This defined the requirement for a fast-climbing interceptor, capable of climbing to 50,000 feet in six minutes following launch. BuAer had already initiated a program with Douglas for a delta-winged interceptor, its Model 571, with the emphasis on climb rate and altitude capability. It subsequently held a competition for this requirement, with the Douglas design being in effect, the baseline. (It became the F4D Skyray). Mission equipment was to include 1.) a search and track radar for all-weather acquisition and attack of incoming enemy bombers and 2.) unguided rockets as its primary armament.

As usual, Vought did design studies with different planforms and engine combinations, including one with two P&W J42 engines and a bicycle landing gear and another with two Westinghouse J40s in a tailless design with a single fuselage-mounted fin. Single-engine J40 designs predominated, however, since the Navy seemed to be favoring it for its interceptor program.

Vought prepared proposals for both a twin J46-powered variant, the V-362A, and one powered by a J40 engine, V-362B, powered by that engine. Unlike the twin-engine V-362A, the J40 tailless variant proposed took full advantage of the relatively high stall speed, 105 knots, permitted by the Outline Specification for the landing gross weight. The wing aspect ratio was only 2.5, the sweep was increased to 45°, and its thickness ratio (parallel to the plane of symmetry) was reduced to 7.6 percent. Its weight was about the same as the J46-powered airplane. The wing changes meant that it was faster than the twin-engine airplane but the time-to-climb suffered, although at 5.7 minutes it was still predicted to be slightly better than the requirement. The takeoff distance in a 25-knot wind was also higher, 450 feet compared to 280 feet for the twin-engine air-

Above, the BuAer baseline for the OS-113 Interceptor Competition was the Douglas Model 571 based on WW II German research. Douglas received a study contract in June 1947 for preliminary investigation and engineering development up to and including the mockup. It resulted in the F4D Skyray program. (NA 80-G-395917) Below, Vought accomplished predesign studies of almost every potentially feasible configuration and powerplant, including this conventional airplane powered by twin afterburning P&W J42 engines. If nothing else, this effort resulted in comparative performance data that demonstrated that its proposal choice was the best. (Vought History Center) Bottom, the first XF7U was photographed in January 1949 with an intake splitter plate like the one proposed on the V-362 for the OS-113 requirement. It apparently didn't provide the projected benefit since it wasn't incorporated on the production F7U-1. (80-G-397967 cropped)

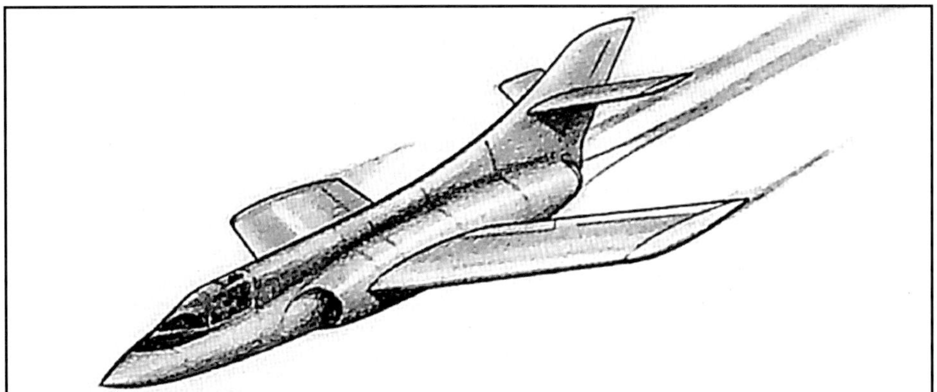

plane.

For some reason, probably to reduce weight and since there was less need for high angle of attack maneuvering, the J40-powered design had a single vertical tail that was mounted on the fuselage rather than the twin fins of the Cutlass layout. In any event, Vought management decided not to propose it because they did "not feel that the proposal of a completely new airplane is justifiable."

On 17 September 1948, Vought hand-delivered the proposal to BuAer for its twin-engine V-362. To provide the increased thrust to meet the time-to-climb requirement, the engines were now Westinghouse J46s with 4,080 lbs thrust each at the military rating and 6,100 lbs thrust for combat. The wing planform was identical to the production F7U-1's but the inboard leading edge slats were now the same chord as the outboard, which somewhat increased the lift. The fuel cells in the outboard wing were deleted since the interceptor didn't need as much internal fuel.

The stub fins beneath the wing were thinner for less drag at transonic speeds. In order to continue to use them as wheel wells, the landing gear tires were to be solid rubber so as to be as narrow as possible. In order to provide more volume in the nose for avionics, a solid rubber tire was also used on the nose landing gear along with a forward retracting strut, since it was slightly shorter than the aft retracting one for the same stance. One design feature that was carried over to the production F7U-1 was the two-position main landing gear devised to eliminate the potential for tail bumping on touchdown at an angle of attack for the slowest landing speed.

An AN/APS-25 search and track radar was incorporated. Since the mission was to shoot down enemy bombers, the armament consisted of unguided 2.75-inch folding-fin rockets rather than the usual 20mm cannon. These were housed internally in two extensible "baskets" located in the belly of the mid-section of the fuselage, each containing 12 rockets.

With full internal fuel, the gross weight was 18,463 lbs; the combat weight was 16,753 lbs. The time to climb to 50,000 feet with full internal fuel was projected to be 4.7 minutes, significantly less than the six-minute requirement. The stall speed was also 18 knots lower than the maximum allowed, and the maximum speeds higher by 30 knots at sea level and 10 knots at 50,000 feet.

Vought noted in its proposal that "While the delta-wing or other extreme planform versions offer some attractive high speed possibilities, it is felt that knowledge of the low speed characteristics of such planforms is not sufficient at this time to permit prediction of adequate carrier landing and takeoff characteristics."

Below, for the OS-113 competition, Vought settled on and proposed a fighter with the F7U-1's wing and fins combined with a new fuselage configured for the interceptor mission. (Author)

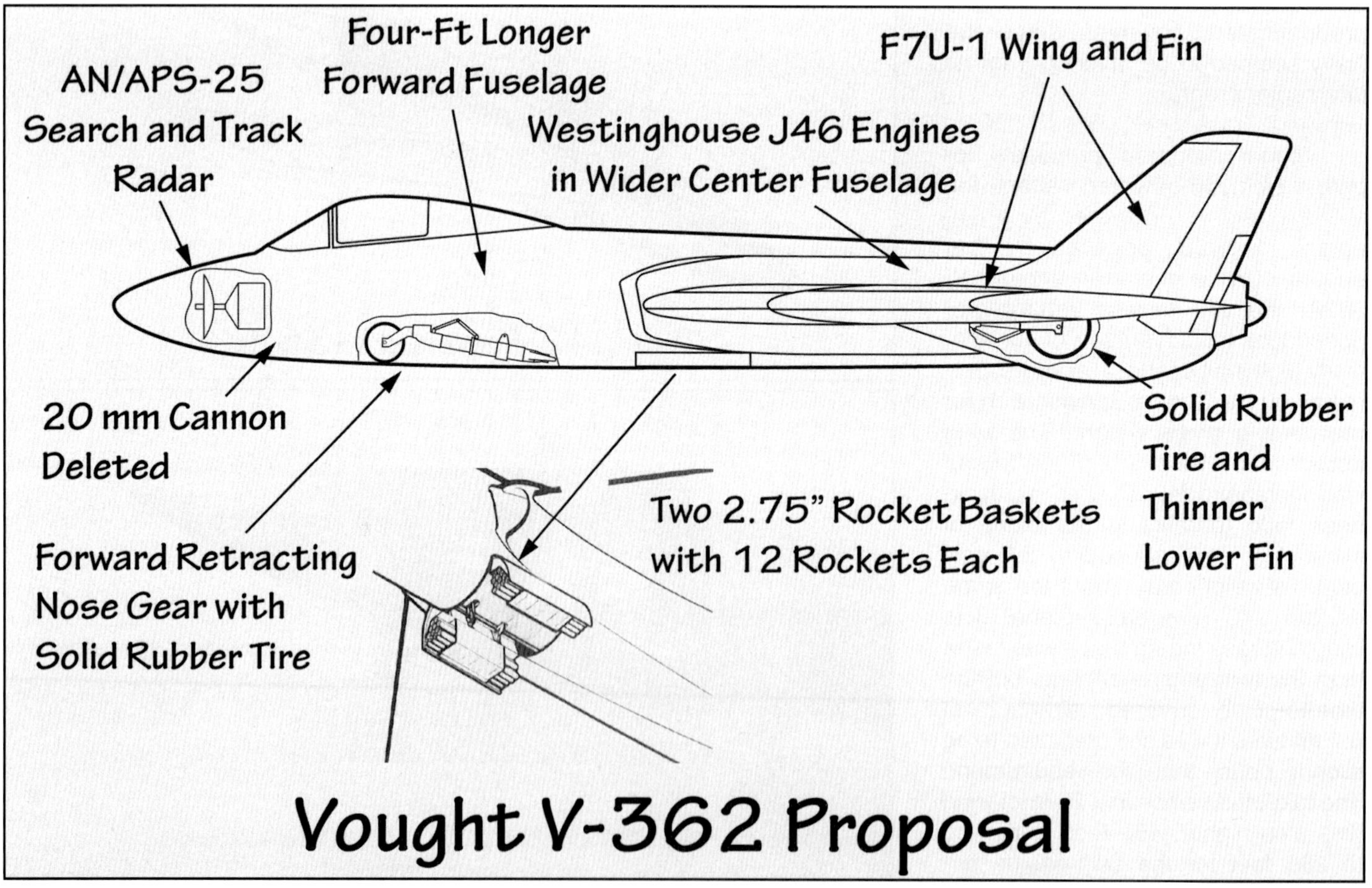

This was a not very subtle commentary on the Douglas D-571 program. Vought reiterated its conviction that conventional planforms were inferior to tailless, particularly for maneuverability at 50,000 feet: "A careful study of tailed and tailless airplanes showed that both were of approximately the same size, but the tailless configuration showed slight performance superiority."

McDonnell, North American, Grumman, and even Douglas also submitted proposals for the OS-113 requirement. According to the BuAer summary evaluation and recommendation dated 24 November 1948, the two Douglas proposals, one identical to the quasi-delta D-571 for which Douglas already had a study contract and the other conventional, were both outstanding. Since the Navy planned to contract for the D-571 as the XF4D, "a more conventional arrangement is most desirable for insurance," and "it was not considered to the best interests of the Navy to have both the conventional and unconventional designs in the same plant at the same time," both Douglas proposals and the Vought tailless design were eliminated from consideration. McDonnell's Model 58 proposal was well-regarded for its speed and climb performance, and its maneuverability at altitude was second only to the Douglas delta design. It was therefore selected for a Phase I contract and designated F3H.

Navy Preliminary Evaluation

The first and only formal F7U-1 Navy Preliminary Evaluation (NPE) was accomplished at Ardmore in June 1949. Five Navy pilots flew BuNo 122474 for a total of 14 hours in 21 flights. As a result of the NPE comments and Vought flight testing to that point, several changes were to be incorporated: the vertical fins were to be increased in area and aspect ratio; yaw rate damping was to be added; the control feel system was to be revised to move the pitch control feel units to the wings; and the wing speed brakes were to be modified.

The original speed brakes were mounted on the trailing edge of the wing inboard of the fins. They proved unacceptable because when opened, they created excessive buffet loads on the fins. The fins were therefore reduced from a total of 40 square feet to 28.5 square feet by narrowing their span and making the hinge line perpendicular to the centerline of the aircraft.

The original design requirement for the speed brakes was that the airplane would not exceed "the Mach number critical for control" in the first

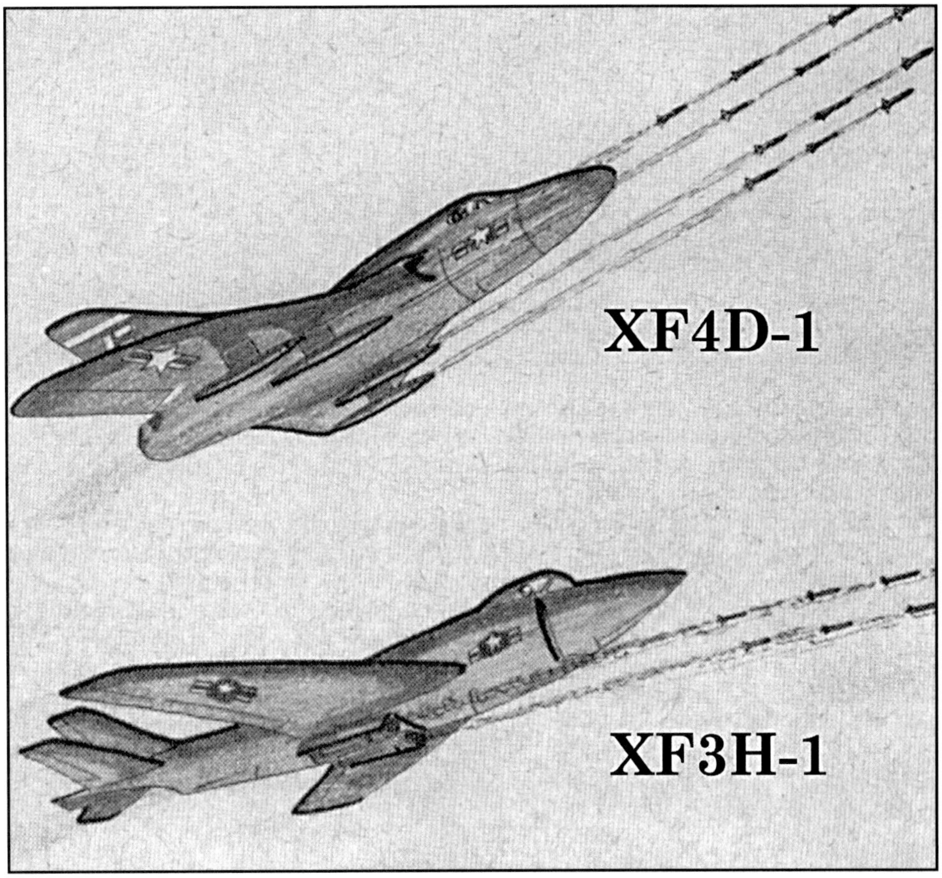

Above, the McDonnell Model 58 proposal was selected for a development contract award along with the Douglas Model 571. As depicted in an October 1949 Naval Aviation Confidential Bulletin, the XF4D Skyray was to carry pods of 2.75-inch unguided rockets as armament whereas the McDonnell XF3H Demon would carry its rockets in a retractable tray. (Author) Below, the original wing-mounted speed brakes had to be reduced in size to eliminate buffet on the fins. Two were therefore added to the underside of the engine inlets to restore the original area as displayed on this production F7U-1, BuNo 124422. (Jay Miller collection)

Above, the F7U-1 prototypes were frequently modified to evaluate design changes that might resolve performance or stability and control issues. In this case, XF7U-1 474 is to be flown with a long dorsal strake and a short ventral one to evaluate this configuration as a fix for a directional stability shortcoming. (NA 80-GK-8698) Below, as a result of the NPE conclusion that the visibility on approach was unacceptable, Vought engineers prepared a polar vision study. It clearly indicated that the 12-degree over-the-nose visibility requirement wasn't valid for an airplane approaching at the 19-degree nose-up attitude of the F7U. It needed to be twice that. (Author) Bottom, drooping the nose, cutting back canopy structure, and changing the pilot's eye position were solutions evaluated to improve the visibility over the nose to be acceptable for carrier landings. (Author)

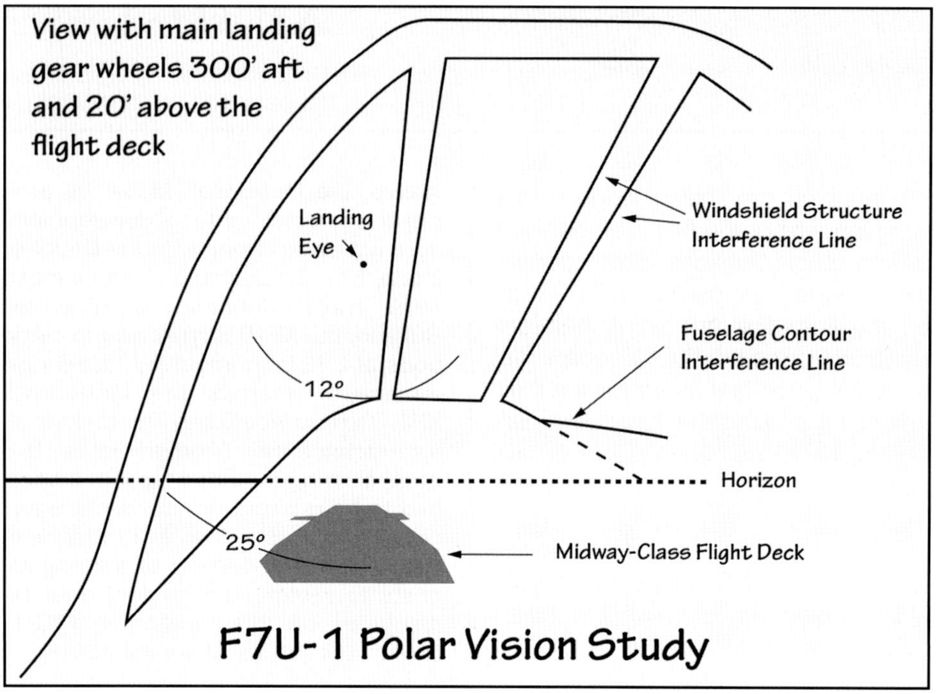

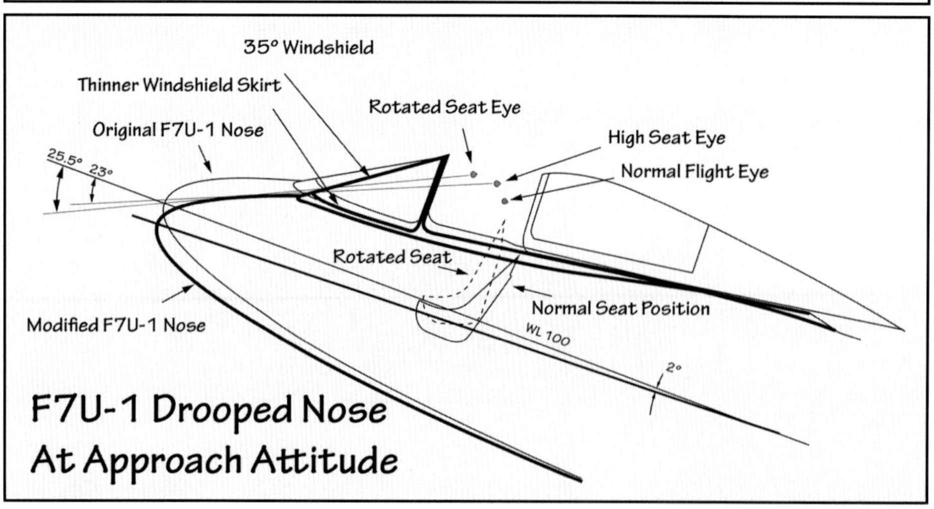

15,000 feet after a vertical dive was begun, with the brakes extended, at 40,000 feet. They also had to slow the aircraft to about 350 knots quickly following a boost failure at high speed, at which point the flight control loads were not excessive for the pilot. The reduced area had to be replaced.

Vought therefore added two speed brakes with a combined area of 11.2 square feet under the fuselage aft of the engine inlets. In addition, air bottles for emergency extension were added so that a complete hydraulic system failure at high speed would not prove catastrophic. At some point, the speed brake actuation mechanism was modified so that the upper wing speed brake opened to a greater angle than the lower to insure a nose-up pitch change when they were actuated, which might be important when needed for a boost-failure deceleration.

An emergency electric-powered hydraulic pump was also added to provide roll control if the basic utility hydraulic system failed. It was originally a ¼ gallon-per-minute flow; this was subsequently increased to ½ gallon-per-minute continuous.

The NPE pilots considered the approach vision to be unacceptable: "it is difficult to consistently obtain impact sinking speeds below design limits in shipboard operation," As a result Vought engineers began examining design changes to improve the approach visibility. These included a raised seat, canopy changes, and the tilting of the forward fuselage downward a few degrees.

XF7U No 1 Bites the Dust

The Navy soon discovered that none of its first-generation jets carried enough fuel. The solution was the addition of tip tanks and in some cases, a fuselage stretch to add internal fuel. Tip tanks were generally a straightforward addition and in the case of straight-wing jets, they were on the center-of-gravity, creating no cg shift as fuel was consumed.

Tip tanks on swept-wing jets

were a more difficult proposition. Centering them on the tips caused a forward movement of the cg as their fuel was consumed, requiring more elevator control power for approach and landing. Mounting them so they extended well forward of the tips and close to the takeoff center-of-gravity introduced aeroelastic design requirements for the wing structure that were heavy and might fail to forestall torsional twisting under all flight conditions.

The minimal capability of the ailavators to handle much difference in cg dictated at least a look at tip-mounted fuel tanks. NACA was enlisted in this effort and accomplished wind tunnel tests.

The selected configuration was a pylon mounted on the wing inboard of the landing gear with the tank center-of-gravity directly under the most forward airplane cg. The pylon was huge, presumably to minimize interference drag between the tank and the fuselage and wing. The tank had a capacity of 250 gallons and according to the Vought drawing, was sized proportionally to the Douglas low-drag shape. No pictures were found of the external pylons installed on the airplane. A takeoff crash in September 1949 may have been the reason.

During initial flight test of the XF7U, Vought had determined that the yaw angle of the airplane versus rudder angle was greater than desirable. Various changes were considered to correct this: a small reduction in the size of the rudders, reduction in the amount of rudder deflection from 25° to 15°, and a differential rudder system. The last would result in only one rudder deflecting to the side that the pilot wanted to yaw toward. In other words, when the pilot pushed on the right rudder pedal to yaw right, only the right rudder would deflect in response.

BuAer requested that NACA Langley evaluate the suggested modifications to rudder control to ensure that spin recovery was not affected by any of the proposed changes. Tests of a .057 model in the Langley 20-foot free-spinning tunnel indicated that the reduction in rudder deflection and/or reducing the size of the rudders "did not seriously alter the spin or recovery characteristics of the model." Some benefit was expected from the differential rudder system in some spin situations because it resulted in the rudder on the "inside" of the spin not offsetting the beneficial yaw moment produced by the rudder on the "outside" of the spin by producing pro-spin drag. However, the test results did not show a measurable impact. NACA's only cautions concerned the potential for high rudder loads that would adversely affect the recovery and the need to delay forward stick for ½ to 1 turn after rudder input.

Once the spin recovery capability of the differential rudder system was predicted to be acceptable, it needed to be flight tested. The first XF7U-1,

Above, tip tanks were rare on conventional swept-wing airplanes because using the fuel would cause the center-of-gravity to shift forward significantly. The shortcoming was even worse for a tailless configuration with its smaller acceptable cg range. However, NACA Langley conducted wind tunnel tests of a wide variety of external tank options, including the tip-mounted tanks shown here. (Tony Landis) Below, Vought settled on external tanks mounted on pylons inboard of the fins. The reason for the extremely deep pylon was probably to avoid interference drag between the fuel tank and the wing and/or engine nacelle. (Author)

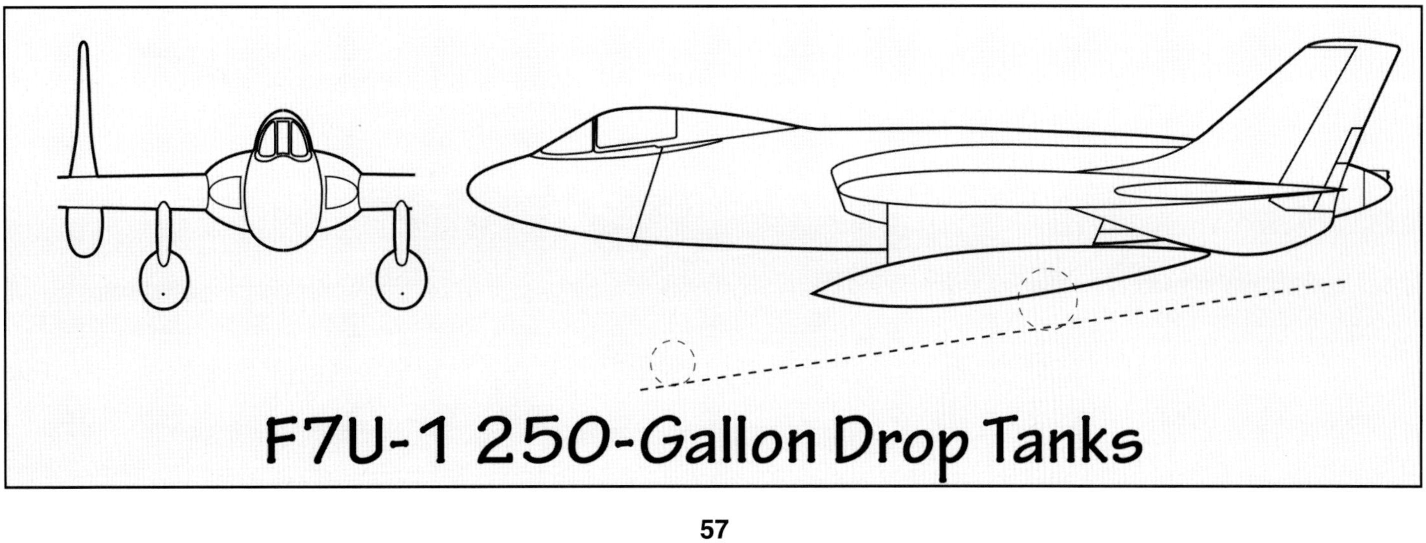

BuNo 122472 was modified accordingly in early September 1949 and several flights were made to evaluate the change. These were satisfactory so envelope expansion continued with asymmetric loading. Which is why Paul Thayer began a takeoff roll on 29 September 1949 at Ardmore with a simulated 150-gallon external fuel tank on the right-hand wing pylon, which was located inboard of the main landing gear. He was unable to keep from rolling too far to the left immediately after liftoff, assuming that the report is correct about which wingtip hit the ground first. In any event, the airplane cartwheeled to destruction, with the forward fuselage breaking off and winding up against the leading edge of a wing. Thayer was not seriously hurt, merely bruised, cut, and shaken-up.

A formal report on the crash could not be found. According to a summary in a Navy monthly report, the airplane yawed excessively to the left shortly after becoming airborne, struck the ground, and cart-wheeled. "From photo observer data it appears that the pilot applied excessive right-rudder pedal forces but obtained only 6 degrees right-rudder deflection." (Excessive in this case didn't mean more than desired, but rather a lot more than should have been required: six degrees is far less than the rudder should have deflected.)

One might think that a heavier load on the right wing would result in a right roll developing at lift-off with the right wing tip hitting the ground, not the left. One scenario that fits all the stated facts (the amount, if any, of crosswind is not available) goes like this: During the takeoff roll, Thayer was applying left rudder due to the higher drag of the right tire caused by the greater weight on the right wing. Nearing liftoff, Thayer was leading with left stick in expectation of a right roll due to the asymmetric external load. At liftoff, the left rudder—no longer required to counteract the higher drag of the right tire—caused the airplane to yaw to the left, increasing the lift on the right wing and decreasing the lift on the left wing. The roll to the left caused by the left yaw was initially exacerbated by the left aileron input intended to account for the asymmetric load. At low speeds, rudder is more effective

Above, XF7U-1 BuNo 122472 with BuNo 122474 (474's main gear is visible behind 472) probably during the ferry from Patuxent River to Texas. (Edmund T. Galbraith via Menard) Below, XF7U-1 122472 taxiing. 472 was unique in that it was the only -1 Cutlass to receive intake splitter plates. (Ginter collection)

BuNo 122474 - Original XF7U-1 Tail

Above and below, the original tail fins (above) on XF7U-1 BuNo 122474 were modified with a redesigned vertical fin and rudder (below) for flight evaluation. It was also used for evaluation of the nose-wheel pre-rotation feature designed to reduce kickback loads on the nose-gear strut and actuator. (Author)

than aileron at producing roll. However, either the differential rudder system limited the rudder control power available to Thayer or the rudder had jammed, giving him only six degrees of deflection.

In any event, there is no more mention of the incorporation of a differential rudder system or apparently few if any flights with external fuel tanks.

With the loss of 122472, all of the Cutlass flight test development effort had to be borne by the sole remaining XF7U-1, 122474, until the first production F7U-1 flew, which was not to be for another five months. At the time of the crash, 474 was laid up for configuration modifications. It was therefore not available for flight test until 17 October. For almost three weeks there was no F7U on flight status. In light of the number of problems that Vought was working to resolve, this had a direct impact on the schedule.

Vought ended the year with one step forward, however. When the redesign to the longitudinal flight control system did not improve the handling qualities, a flight was made with the hydraulic control components removed or disconnected, using only the direct spring-tab control of the ailavators in manual mode. The pilot reported that in testing up to Mach 0.8, the longitudinal controllability of the airplane compared favorably with other airplanes: there was no porpoising and he could maneuver with precise control, demonstrating that the control sensitivity was not an inherent aerodynamic condition but one caused by the hydraulic flight control system itself. Capt. Trapnell, now the commanding officer of NATC, flew 474 on 22 November 1949 in that configuration and concurred.

Over the next five months, however, the only F7U available spent more time in layup than on flight status, as various modifications were installed and then checked out. During the first layup, the redesigned tail fins and rudders were installed along with a "master-slave" hydraulic linkage between the pilot and the flight control valve.

BuNo 122474 - Production F7U-1 Tail

Oriole

As a result of World War II experience and fleet air defense studies/trials, air-to-air guided missiles were determined to be far more likely to down attacking jet bombers aircraft than attempting to close and fire at them with 20mm cannon. The Navy therefore embarked on a plan to develop three different missiles for short, medium, and long-range use. Unguided rockets were to be used in the interim; they weren't very accurate but just one hit was likely to bring down the bomber.

In 1947, Martin received a contract from the Navy for development of an on-board radar for the medium-range air-to-air missile, project Oriole. The proposed missile was 11 inches in diameter and 11.5 feet long. It was to weigh almost 400 lbs, including the radar, a 50-lb warhead, an auxiliary power unit for electrical and hydraulic power, and the requisite electronics to track the target and guide the missile.

In July 1949, the Air-Launched Branch of the Pilotless Aircraft of BuAer asked Division Design Division to evaluate the suitability of various aircraft to carry the Oriole. The analysis was simplistic but comprehensive as to prospective platforms, providing assessments of the F7U-1, AJ-1, XA2J-1, F3D-1, XF4D-1, XF3H-1 and XF10F-1. The major considerations were gross weight and center-of-gravity impact, provision of an acceptable field of view for the missile's radar, and altitude/airspeed capability. BuAer assumed a desired launch speed of Mach 0.8 and a minimum launch speed of Mach 0.65.

The F7U-1 was the only one of the aircraft studied that required a forward fuselage change to accommodate a modified AN/APS-25 radar with a 24-inch scanner. The BuAer engineers were either unaware of the Vought design for wing-mounted pylons inboard of the fins or rejected it. In any event, they concluded that the F7U-1 could only carry two Orioles, mounted on pylons located on the engine nacelle. A wing-mounted location was considered unacceptable due to 1) the presence of fuel

Below, an in-house Navy study of prospective Oriole missile platforms concluded that the F7U-1 could carry a maximum of two because of the perceived need to carry the missiles on the center-of-gravity and unacceptability of trying to attach pylons below fuel tanks. The field of view of the missile radar from the pylon location chosen was partially blocked. The engineers were apparently unaware of the Vought design for a pylon on the inboard wing. (Author)

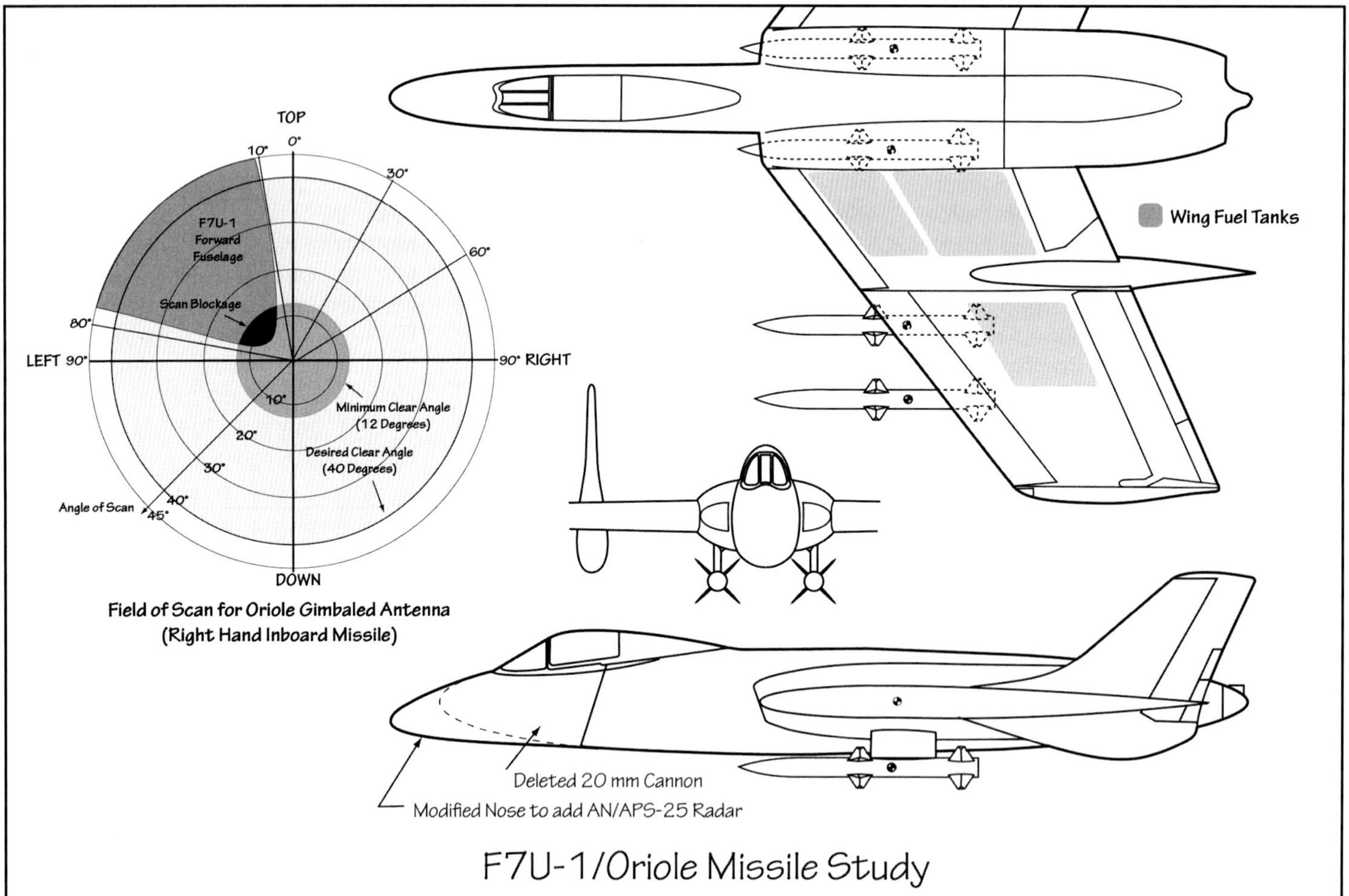

F7U-1/Oriole Missile Study
Bureau of Aeronautics
December 1949

tanks in the inboard wing which precluded the attachment of the pylon and systems interface with the missile and 2) the size and configuration of a pylon required to carry the missile close enough to the aircraft's center-of-gravity. This location, however, resulted in a restricted field of view for the missile's radar.

The weight of the longer nose, requisite radar, missile pylons, missiles, etc., required the removal of the 20 mm cannon armament, associated structure, and ammunition in order to keep the gross weight within limits. These removals totaled 1,355 pounds including the deletion of 138 pounds of ballast required on the existing design for a center-of-gravity between the recommended limits of 16.2 and 19.5 percent of MAC for all possible loading conditions.

Performance was projected to be excellent based on existing predictions. The F7U-1 was estimated to be able to carry two Orioles at speeds above Mach 0.8 at any altitude up to 40,000 feet by using afterburning.

The study concluded that the AJ-1 and XA2J were probably inadequate launch platforms from a speed standpoint and the F7U-1, given the restrictions that had been placed on the pylon location, did not provide an adequate field of view for the Oriole radar. The F3D was marginally acceptable as to speed but could carry four missiles. Of the remaining fighters, the XF10F could also carry four missiles; the XF3H, three; and the XF4D-1 only two like the F7U-1. As a result, the F3D was considered to be the most suitable for test purposes since providing swiveling mission pylons on the XF10F's variable-sweep wing would require additional cost and time for development.

As it happened, the study conclusions about the F3D were validated by subsequent events. It was used as the workhorse for the air-to-air missile development programs; Oriole missile firings were accomplished in 1952 from an F3D-1. It was also the Navy's first guided-missile-armed fighter, armed with the simpler Sparrow I. The Oriole program itself proved to be too ambitious for the technology of the time. The simpler Sparrow I proved to be marginally acceptable and became operational, first on the F3D-2M and subsequently on the F7U-3M, which proved not to have a cg problem with four missiles after all.

As for the other aircraft in the study, the XF10F program floundered, although the variable-sweep wing was not the cause: the ill-fated Westinghouse J40 engine and Grumman's attempt to avoid hydraulic-powered flight controls did it in. The most successful missile platform operationally proved to be the F3H-2 Demon. Douglas unsuccessfully proposed air-to-air guided missile capability for the F4D but did receive a contract for the radar-equipped Sparrow II, which encountered the kind of technical difficulties that killed the Oriole program, to be carried by the F5D. Both missile and airplane were subsequently cancelled, albeit for different reasons.

Above, a VX-4 F3D-2M with four Sparrow I missiles launches from the USS Hancock on 30 June 1954. The F3D-2M was the first guided missile fighter but it did not deploy to the fleet. (National Archives) Below, as it happened, there was no problem with either the location from a cg standpoint or mounting of missile pylons on the F7U-3M Cutlass, which became the first carrier-based fighter to deploy to the fleet with the Sparrow I missile. (Vought Heritage Center)

Above, the Navy reviewed the production F7U-1 cockpit mockup on 6 and 7 October 1948. The ejection seat now incorporated the Martin-Baker face-curtain ejection-activation concept. (USN, National Archives) Below, the changes provided to Vought to the October 1948 F7U-1 review were incorporated in the F7U-1 cockpit mockup for the lighting review that was accomplished in December. At this point, a chart board was still a requirement; it would be deleted from the F7U-1 in favor of a small "kneeboard" that was part of the pilot's personal equipment. (USN, National Archives)

F7U-1 Production Contract

It was not unusual for BuAer to order a small quantity of production aircraft even before the first flight of the prototype. It was understood that there would be changes required as a result of flight test that would have to be incorporated in the production aircraft as they were being built or even after they were completed and first flown. These early production aircraft would be used for company flight test as well as Navy evaluation. If all were going well, a second, larger, batch of airplanes would be ordered a year after the first, and so forth on an annual basis.

The first batch would provide for contractor flight test and Navy Board of Inspection and Survey (BIS) Trials: flight test, electronics test, tactical test, armament test, and service test. At the time, these took 12 to 18 months to complete. After the BIS airplanes were delivered, ongoing production would be delivered to development squadrons for further evaluation. If the XF7U-1 development progress and performance were meeting expectations in July 1949, several months after first flight, a follow-on order for a greater quantity would be forthcoming for deliveries in 1951. The F7Us in the second production batch would be issued to fleet squadrons, which would deploy after a suitable period of training and qualification.

It was understood and accepted that the modifications would be required to the first batch as a result of flight test with the XF7Us and that updates to the second batch were also likely based on changes resulting from test of the first. This approach entailed some risk of higher cost and, worst case, the embarrassment of having built airplanes that were determined to be unsuitable for service use. It usually resulted, however, in the quickest introduction of state-of-the-art aircraft in the fleet to maintain parity if not superiority with prospective enemy air forces.

In order to minimize the risk of investing in an aircraft that would turn out to not meet contract requirements, the Navy had established periodic milestones for program review. Prior to first flight, these included mockup and engineering design reviews. After the contractor had expanded the flight envelope somewhat, Navy pilots flew the airplane for a formal Navy Preliminary Evaluation, followed by formal and informal flight evaluations every so often.

The Navy also had a practice of ordering new or improved designs for the same mission requirements from two or more different manufacturers. If one failed, the other would hopefully prove to be satisfactory. If two proved acceptable, the Navy might keep both in production as they did with the F6F Hellcat and F4U Corsair during the war, maintaining a competition for future production.

In this case, the Navy was particularly eager to regain parity with land-based fighter airplanes that had swept wings and higher performance. So in spite of the weight increase, on 18 May 1948 BuAer requested authority from OpNav to procure 20 F7U-1s along with 35 F6U-1s. Due to fiscal year funds availability, the F6U quantity was subsequently reduced to 33 airplanes and the F7U, 19. The deleted airplanes were to be reinstated on 1 October 1948 but only the F6Us were. The Bureau Numbers assigned to the F7Us were 124415 through 124433. The first production F7U of this order was to be delivered in December 1949 and the last, June 1950 as production deliveries ramped up to four per month. Vought received a letter of intent for F7U-1 production from BuAer in July 1948.

The F7U-1 cockpit mockup was reviewed on 6 and 7 October 1948 at Stratford and accepted with the understanding that approved changes would be incorporated.

The change to the Mk 6 Mod 0 Fire Control System for the F7U-1 from the Mk 7 specified for the XF7U-1 along with the addition of a radio altimeter and radio compass to the right hand console, relocation of the engine RPM indicators to the upper left-hand side of the instrument panel, and the deletion of chart board stowage resulted in a complete rearrangement of the instrument panel and relocation of the IFF control and lighting control panels.

Carrier-based airplanes up until then had provision for the stowage and support of a large chart board. The chart board was removable so the pilot could take it with him after landing to plan the next flight. The XF7U-1's was typical, a large horizontal slot in the instrument panel that the board could be stowed in and extended from for use in flight.

Instead of the cumbersome board, the Navy had decided to use a small kneeboard chart holder instead (it was actually strapped to the thigh).

The left-hand console was rearranged to accommodate a more

Above, engine control was now simplified with afterburner activation now incorporated with the throttle handles on the left-hand console. (USN, National Archives) Below, all the radios and lighting controls were located on the forward section of the right-hand console. (USN, National Archives)

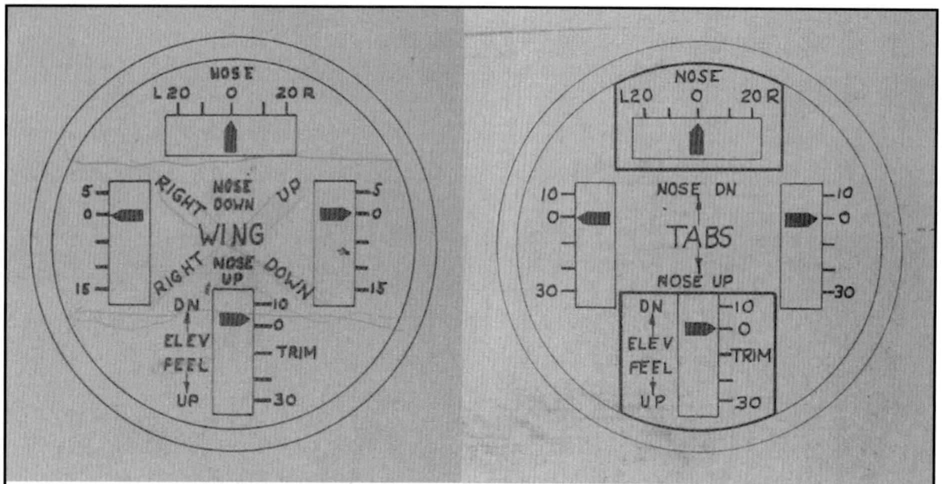

Suggested Approved

F7U-1 Production Trim Indicator

Above, the trim indicator recommended by Vought was slightly modified by the attendees and incorporated. (USN, National Archives) Below, in April 1949, the Navy reviewed a mockup of the XF10F with a variable-incidence wing. There was one more configuration change, to a variable-sweep wing, before detail design could begin in earnest. (Northrop Grumman History Center) Bottom, VF-51 F9F-3 aboard the USS Boxer on 13 September 1949. The Navy began deploying jet fighters aboard carriers in 1949, beginning with VF-51 and VF-52's Grumman F9F-3 Panthers on the west coast. (NA 80-G-407140)

accessible location for the speed-brake control. (A thumb-operated latch was also added to the slat control to preclude inadvertent extension when speed brake actuation was intended.) That resulted in relocation of the afterburner master switches, replacement of the fuel shut-off switches with a smaller rotary switch, and using a push-pull control on the console for hydraulic boost shutoff rather than a lever.

In the original cockpit mockup, Vought had proposed an integrated indicator that displayed the actual position of each of the trim tabs. After discussion, a somewhat simplified design was approved for production.

1949
Navy Fighter Programs Status

In 1949, the Navy was beginning to deploy the first two of its second generation of carrier-based jet fighters, the Grumman F9F-2/3 Panther and the McDonnell F2H-1 Banshee. These were straight-wing day fighters, roughly comparable to the Air Force's Lockheed P-80 and Republic F-84. Both McDonnell and Grumman were also in the process of qualifying improved versions of the Banshee and Panther but neither was to have a swept wing. The big all-weather Douglas F3D Skyknight had a straight wing as well; testing of the prototypes had progressed to the shore-based carrier-suitability evaluation phase but the first production airplane had not yet flown.

The Navy had issued development contracts for two swept-wing all-weather interceptors, the "tailless" Douglas F4D Skyray and the McDonnell F3H Demon. The F4D mockup review was accomplished in March 1949 and the F3H mockup review, in July. Detail design was therefore in progress with first flights scheduled within a year or so. As it turned out, because of Westinghouse J40 engine development problems, the XF4D would not fly until January 1951 even with an interim engine and the XF3H, not until August 1951.

The Navy was also working with

Grumman to define the F10F Jaguar as its first general purpose jet fighter. The program had originated as a Navy-funded study for a swept-wing day-fighter derivative of the F9F-2 after Grumman had insisted on it being designed with a straight wing first. It had metastasized through requirements growth from a gross weight of about 18,000 lbs in early 1948 to 32,000 lbs in late 1949 when detail design finally began. Grumman incorporated a variable-sweep wing in the final configuration iteration in order to keep it carrier compatible.

The first carrier-based jet fighters were unacceptably short-legged even after modification to add more fuel. In order to increase range, they had to be bigger. That meant fewer could be accommodated on an aircraft carrier, which meant that the longer-range fighters had to be multi-purpose as well as all-weather.

A year earlier, BuAer had also decided that all fighter airplanes needed to be equipped with the AN/APS-19 visual-assist radar for use on combat air patrol missions. Adding one to the F7U-1 production airplanes (and air-to-air rocket armament!) had been discussed with Vought as early as May 1948, but both the Navy and Vought agreed that the best approach was to introduce this additional mission capability, and the requisite autopilot for all-weather operation, after production was underway.

OpNav defined this new mission, "General Purpose", in a 27 June 1949 Operational Requirement:

I. Operate in all conditions of weather and visibility
2. Maintain control of air in the vicinity of the task force
3. Conduct fighter sweeps at a distance from the task force
4. Accompany and defend medium range attack aircraft in strikes
5. In keeping with getting the most out of each airplane in the air group on a carrier, the general-purpose fighter was to have "provisions for air-to-ground weapon delivery."

For the general-purpose fighter, speed and rate-of-climb were to be

Above, VF-171 F2H-1 taxis on the USS F.D. Roosevelt on 8 August 1949. The east coast jet fighter deployments began with the McDonnell Banshees, also in 1949. (NMNA) Below, in 1949, the Douglas XF3D-1 SkyKnight was being flown at NATC for evaluation. (NMNA) Bottom, both of the Navy's new interceptors, the Douglas F4D Skyray and the McDonnell F3H Demon, were at the mockup stage in 1949 and more than a year away from first flight. (USN, National Archives)

Above, the McDonnell XF3H-1 Demon, was at the mockup stage in 1949 and more than a year away from first flight. (McDonnell) Below, in 1947, McDonnell proposed a swept-wing derivative of its F2H Banshee that was based on its XF-88 for the Air Force. It was probably the Navy's best chance at deploying a swept-wing jet fighter in time for the Korean War but it passed. (USAF) Bottom, FJ-1 shepherds the swept wing XFJ-2 which did not fly until 27 December 1951. (USN)

secondary to range and endurance. High altitude maneuverability was to be combined with "special aerodynamic features to permit relatively low cruising speed for escort missions." This was clearly a reference to the variable-sweep wing feature on the XF10F, with the low cruising speed dictated by the long-range cruise speed of propeller-driven attack airplanes like the Douglas AD Skyraider.

Since interceptors and the dedicated night fighter were also necessary, the small, high-performance day fighter was in danger of being crowded off the deck and out of BuAer's plans. The Navy had already limited development of an air-superiority fighter in 1949 to only one type, the F7U-1. The Navy didn't lack for alternatives: Vought had suggested a swept-wing variant of the F6U; McDonnell had proposed a swept-wing Banshee that incorporated the wings of its F-88 for the Air Force. BuAer passed on both of those. The F-86 was a logical candidate as well but the Navy chose not to adapt it at the that time. While it was out of character for BuAer to put all its eggs in one basket, it was presumably counting on one of its new high-performance fighter programs to provide an alternative in the event that the F7U proved unworthy.

In late 1948, BuAer was asked to rank its current programs with respect to a requirement to reduce the number. The F7U was considered to be important by the Assistant Chief for Research and Development, Rear Admiral Lonnquist, "because of performance and advanced design." However, the F7U configuration did not compare well to the F10F and the F3H in a December 1948 BuAer review in part due to it having less wing sweep (the F10F's was variable and F3H's was 45°) and thicker wings from a transonic performance standpoint. It also "Lacks tail clearance to achieve maximum lift and suffers known and unknown disadvantages of tailless arrangement." (The tail clearance comment referred to its high angle of incidence on landing).

A Military Requirements memo that was part of that review concluded that the "F7U will need AI (Airborne Intercept) radar to be a completely effective fighter." It was suggested that the AN/APS-25 radar be incorporated along with the J46 engine being developed and increased fuel capacity. (An interim radar configuration was also considered: an AN/APS-19 radar with a 30" dish).

Notwithstanding concerns about the basic aerodynamic suitability of the Cutlass program, it was already flying and the Navy's other high-performance jets were at least a year away from first flight. The designated general-purpose fighter, the Grumman F10F, was still not yet well enough defined for detail design to begin. BuAer therefore kept the J46-powered F7U in mind as a general-purpose fighter alternative.

As a result, by early 1949 Vought initiated another design study effort, the V-366, to define a carrier-based general-purpose fighter powered by either two J46s or one J40, armed with either four 20mm cannon or 24 2.75-inch folding-fin rockets, and equipped with a search and track radar. The configurations went up through at least the letter H, which was a bizarre one with two forward fuselages, one for the pilot and the other for the radar.

Most, however, were minimal variations of the V-362. The V-366D, for example, was virtually identical to the V-362 except that it was armed with four 20-mm cannon mounted under the engine intakes and retained the F7U-1 fuel tanks in the outboard wing panels.

Unfortunately for the Navy's fighter development plans, engine development and qualification problems at Westinghouse caused the Navy to lose considerable time in deploying a generation of high-performance tactical jets. Its interceptors, the F4D and the F3H, were late to the fleet because of the J40 engine debacle. The nonafterburning J34 was acceptable in the F2H Banshee and the F3D Skyknight, but Westinghouse failed badly in adding an afterburner to it for the F7U. As a result, relying only on the Cutlass as its high-performance day fighter instead of having McDonnell or North American develop a swept-wing version of the F2H or the F-86 respectively would delay the presence of one on the carrier decks for a few years.

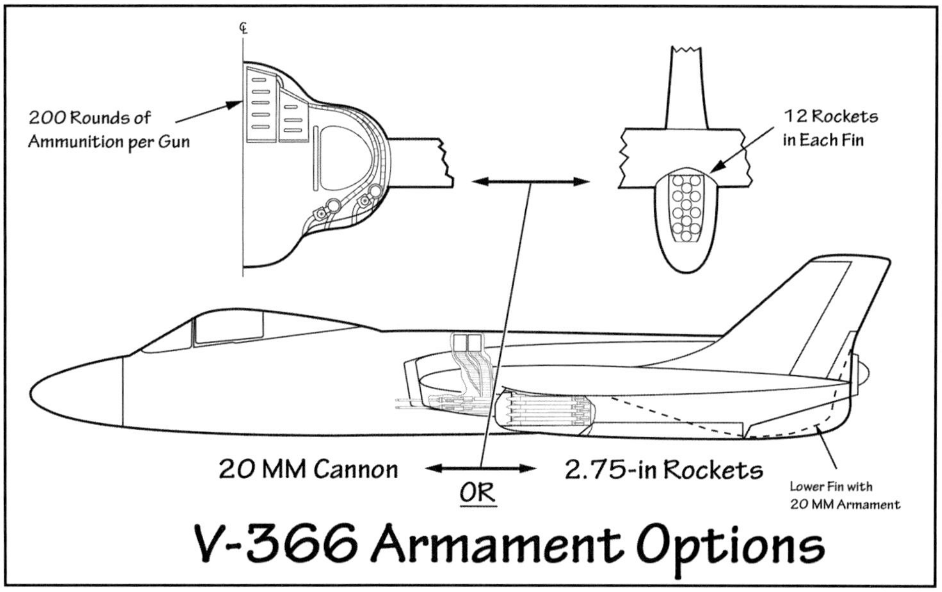

Above, Vought evaluated both 20mm cannon and 2.75" rocket armament for the V-366. Unlike the V-362, the rockets were carried in extensions of the sub fins instead of in a retractable basket in its belly. (Author) Below, One of the V-366 design studies had the objective of shortening the overall length, a critical metric for a carrier-based airplane. It housed the pilot in one nacelle and the radar in the other. The length reduction was significant but it undoubtedly failed the giggle test, at BuAer if not at Vought. (Author)

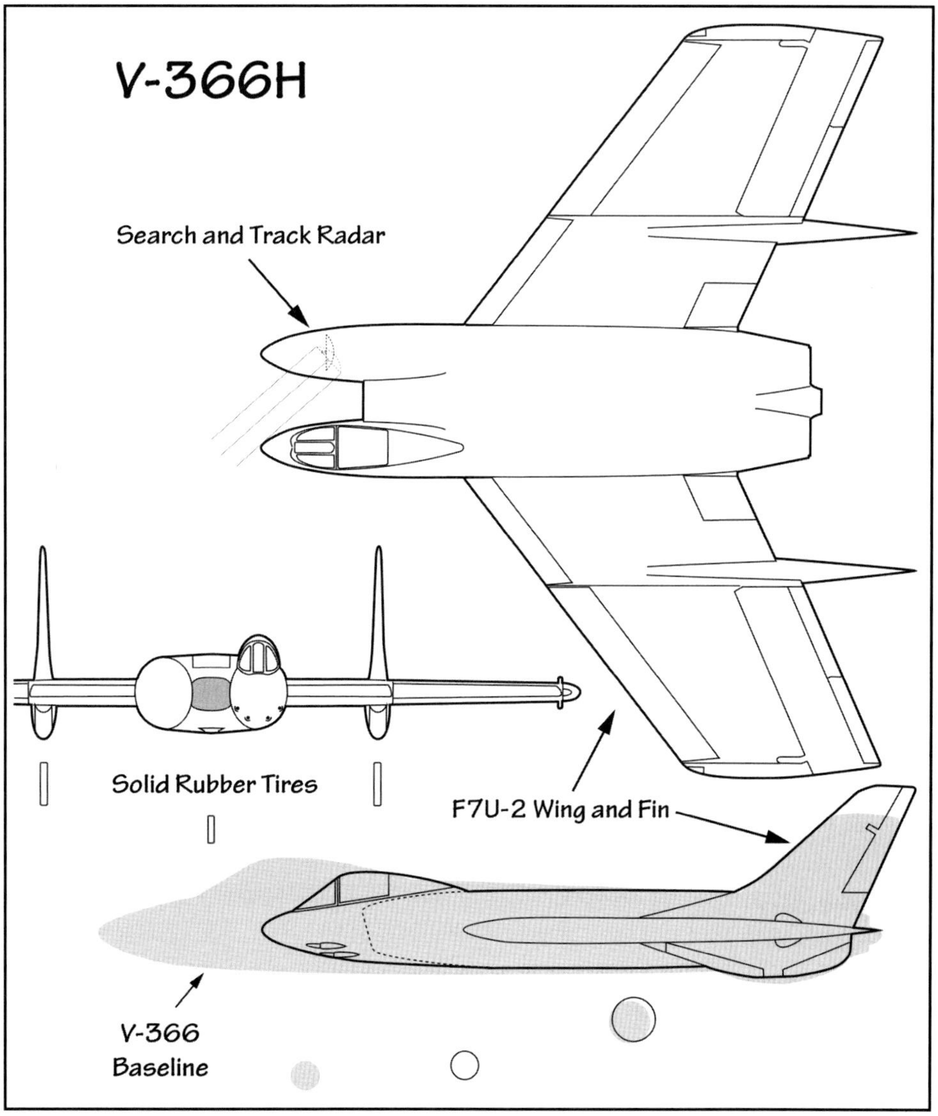

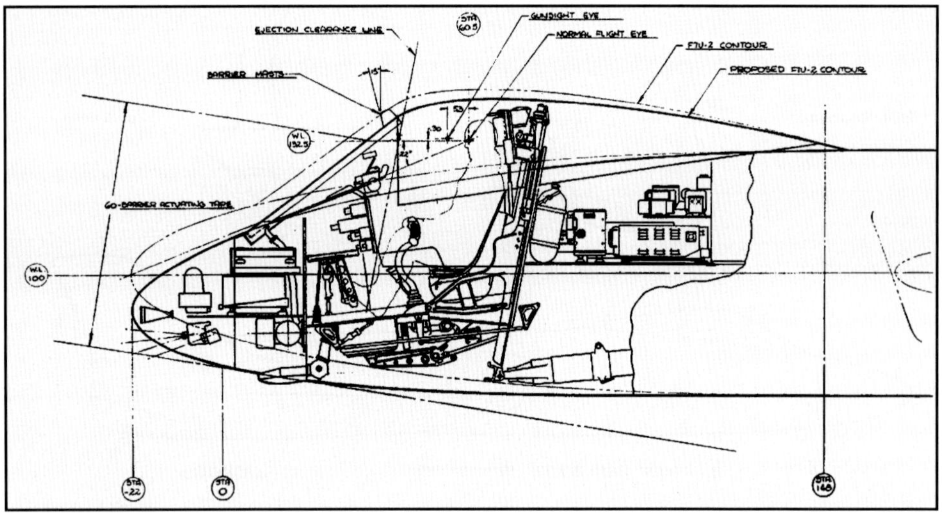

Above, changes to the F7U-2 forward fuselage were evaluated to improve the visibility over and to the side of the nose for carrier landings, including raising the canopy and drooping the nose. (Vought Historical Center) Below, although no drawings could be found of the F7U-2 itself, a notional configuration could be deduced using other drawings and information. However, the likelihood of a J34 thrust increase was significantly less than the certainty of the -2's weight increase. (Author)

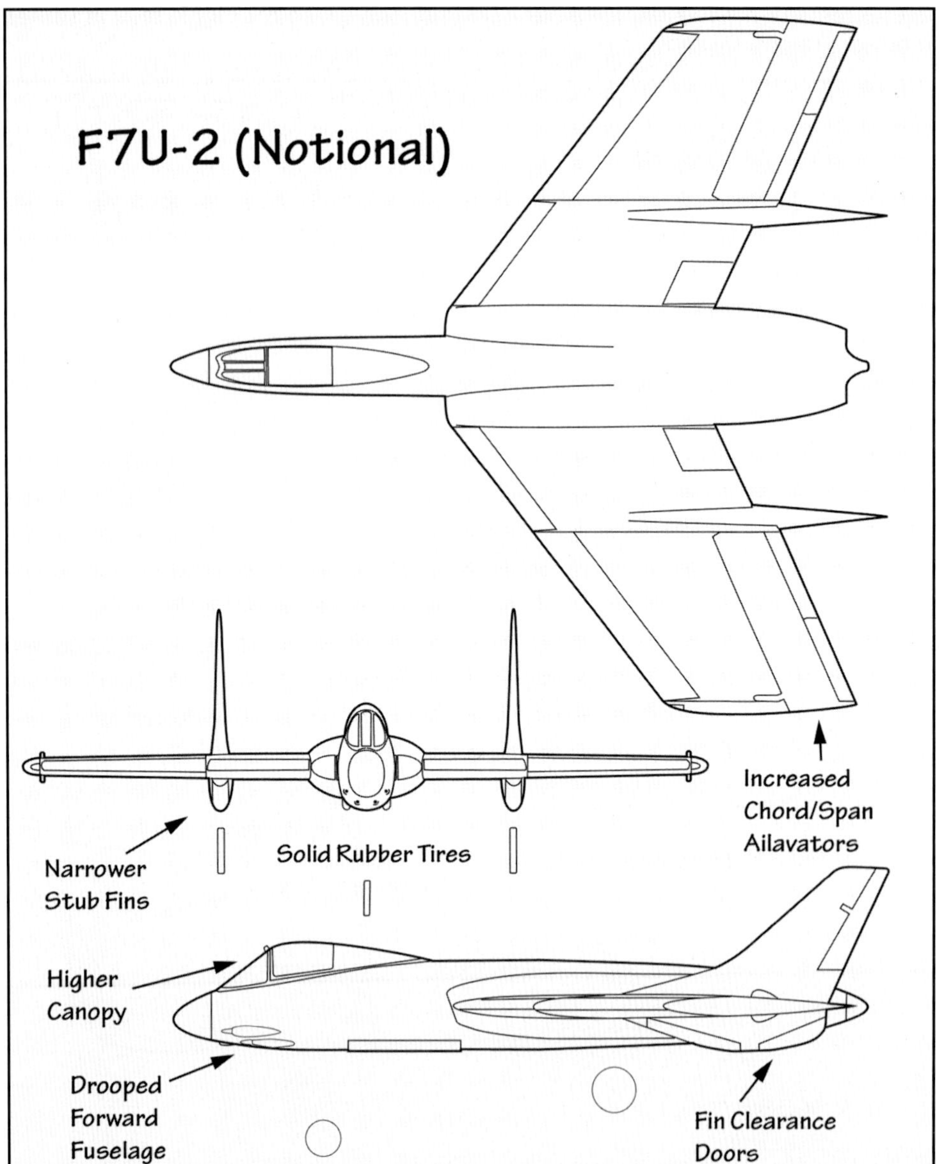

Transition to F7U-2/3

In early 1949, Vought was expecting a July release of an FY1950 order for 96 production F7U-1s. However, the F7U-1's empty weight had continued to increase (another 1,000 lbs since late 1948) and BuAer finally decided it was inadequate from a structural strength standpoint for the flight and landing loads of an operational airplane. Nevertheless, the Navy still wanted to deploy a high-performance day fighter as soon as possible and did not have any nearer-term alternatives. (There was also a need for an interceptor to fill-in until either the F4D or the F3H became deployable.) As a result, BuAer was receptive to Vought's April 1949 proposal for a structurally beefed-up F7U design, with the first one to be delivered in September 1950. A request for the F7U-2 designation was made on 28 July 1949.

The -2 was to have a "strengthened wing and fuselage, a structural change in the aft fuselage section lines to permit (simplified) stainless steel fabrication, and changes to the cockpit canopy and windshield to improve pilot's vision for carrier approach and also sidewards and downwards." The structural beef-up would restore the flight, carrier landing (sink rate, arresting, barrier crash), and takeoff catapult strength compliance but it would otherwise be basically the same. At the time, a "J34-WE-42" engine was planned for F7U-2 production. (This was not the same -42 that Westinghouse subsequently produced as the final F7U-1 production engine.) It's likely that a thrust increase was hoped for but BuAer accepted that the -2 would suffer "some sacrifice of speed and endurance" due to the weight increase without a corresponding thrust increase.

F7U-2 production wasn't a sure thing, however. The Chief of Naval Operations (CNO) had been informally made aware of the F7U flight deficiencies. As a result, his office formally requested BuAer to reevaluate the F7U and cancel or postpone the procurement if "the F7U cannot be made acceptable within the period that 1950 procurement funds must be committed." BuAer responded that the F7U represented a marked step forward in performance over the F2H and F9F and should be continued. Somebody in the Navy or at Vought got the following item in the April 28th 1949 issue of the British aviation journal, *Flight*: "From America it is unofficially reported that the world's record of 670.983, set up by a North American F-86, has been exceeded by the tailless Chance-Vought Cutlass naval fighter."

As before, eleven of the 19 F7U-1s on order would be used for test and evaluation requirements. The rest were to be assigned to OPDevFor (Operational Development Force) projects for early evaluation of the unique qualities of the Cutlass. The first six F7U-2s would be used for test and evaluation with the seventh and subsequent aircraft assigned to operational squadrons.

In all likelihood, the F7U-2s were only intended to be a steppingstone, in this case to the more capable F7U-3 based on the V-362 proposal and subsequent V-366 design studies. As it happened, the Grumman F10F general-purpose fighter program continued to be a concern to BuAer. Its configuration was still not frozen and a variable-incidence wing was being considered to provide adequate visibility over the nose on approach. Since the Navy was usually not comfortable with not having a backup for one of its fighter requirements, it must have been downright anxious about not having one with both a new engine and a wing that had to be raised for takeoff and landing.

As of August 1949, the Navy was therefore planning to buy two F7U-3s with FY 1950 funding to provide an alternative to the F10F. The -3 was to incorporate an AN/APS-25 radar and additional stores stations and fuel. Because of the increase in weight, it was to be powered by the relatively new 4,000-lb thrust Westinghouse J46 engine, 6,000-lbs with afterburning. The first one was to be completed in October 1951 and the second in November. To fund them, the number of F7U-s would be reduced to 88.

Unfortunately, the magnitude of the F7U-2 changes resulted in more extensive changes to the manufacturing tooling than expected. As a result, fabrication of the first F7U-2 fell behind schedule. (Westinghouse was also having delivery problems with J34 engines.) In August 1949, because of the projected delay in the start of F7U-2 production and to avoid peaks and valleys in its manufacturing manpower requirements that would cause loss of learning and increased unit cost, Vought proposed a production schedule revision. It would stretch out F4U-5N production over a slightly longer period of time; increase the number of F7U-1s to be built from 19 to 29 at roughly the same production rate; and delay the first delivery of the F7U-2 from September 1950 to March 1951.

BuAer accepted the production plan changes proposed, except for the increase in F7U-1 production quantity, and provided Vought with a letter of intent on 23 September 1949 for 88 F7U-2s, BuNos 125322 – 125409, and two F7U-3s, BuNos 125410 and 411. The deliveries were to be June 1951 through January

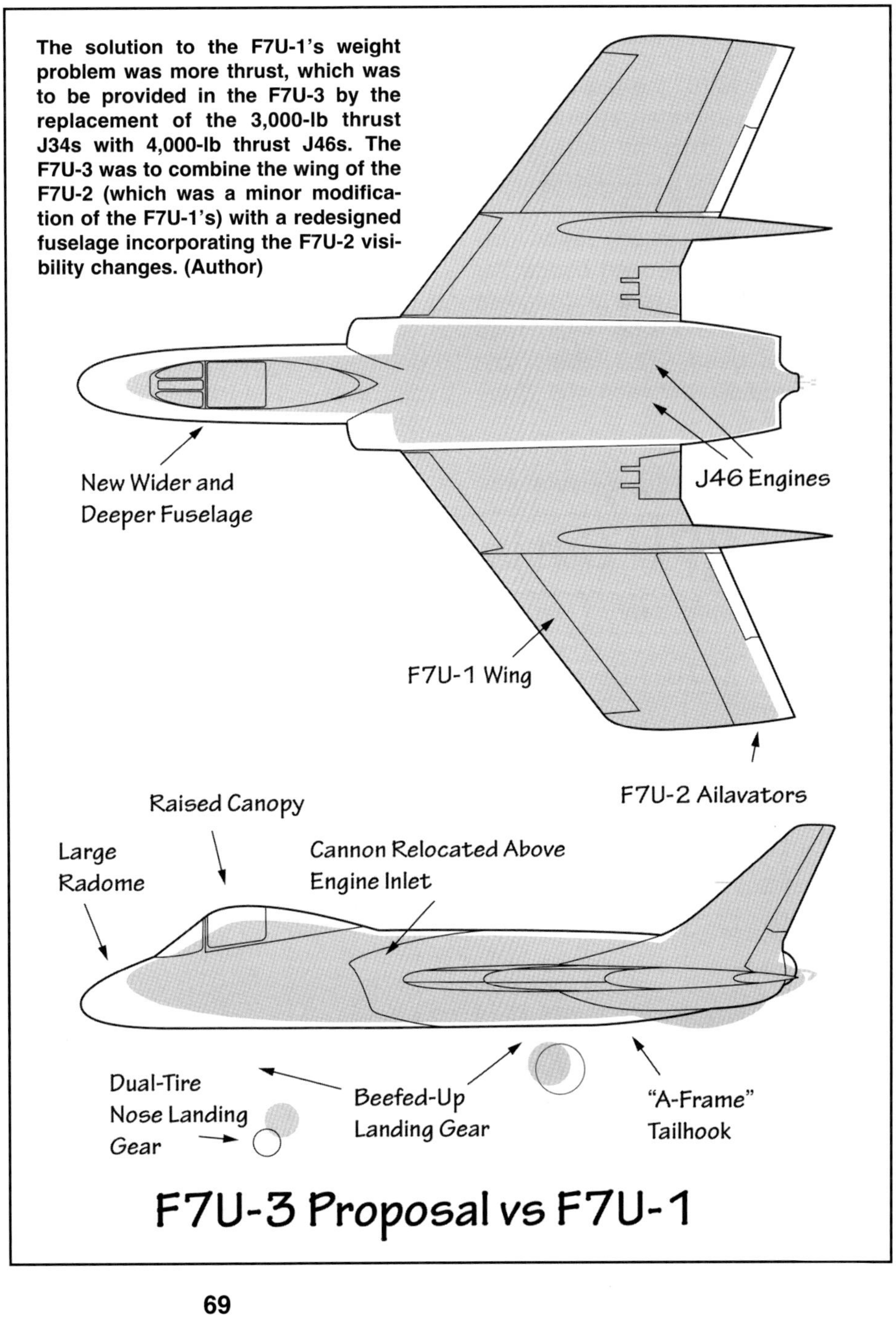

The solution to the F7U-1's weight problem was more thrust, which was to be provided in the F7U-3 by the replacement of the 3,000-lb thrust J34s with 4,000-lb thrust J46s. The F7U-3 was to combine the wing of the F7U-2 (which was a minor modification of the F7U-1's) with a redesigned fuselage incorporating the F7U-2 visibility changes. (Author)

1952. The 19 F7U-1s were still to be completed and delivered. Within a week, Vought also received BuAer's final draft of the F7U-2 detail specification for review and comment.

Addition of a visual-assist radar to the F7U-2 was also a possibility. In testimony before the House Armed Services Committee in October 1949, commenting on the survivability of the B-36, CDR William Martin, then heading up the Pacific Fleet all-weather training squadron, stated that the Navy was considering a night-fighter version of the "F7U-1" in addition to fielding the F3D and the F2H-2N.

The F7U-2 is dead, long live the F7U-3

Prior to 1976, the U.S. government fiscal year began on July 1, six months earlier than the corresponding calendar year, e.g. Fiscal Year 1949 began on 1 July 1950. The president of the United States presented his recommended budget for the upcoming fiscal year to Congress in January. The Fiscal Year 1949 budget had been $13.2 billion, split roughly equally between the three services.

In late 1948, the Secretary of Defense, James V. Forrestal, had recommended a Fiscal Year 1950 Department of Defense budget of $16.9 billion. President Harry S. Truman, determined to continue his practice of balanced budgets, cut that to $14.2 billion in the budget that he submitted to Congress in January 1949. It was not quite as Draconian as it appeared, since it was an increase over FY49's budget. Nevertheless, Forrestal objected and for that and other reasons, Truman replaced him with Louis A. Johnson in March.

In June, just before the fiscal year was to begin, Truman, with Johnson's acquiescence, reduced the DoD FY1950 budget even more, to $13 billion. The battle both within and between the services for operating funds and existing and planned development and production programs escalated. It didn't help that Truman's cold war focus and priority were on nuclear weapons delivered by Air Force, not Navy, airplanes.

As part of implementing the defense budget cutbacks, Johnson moved to adjust the roles of the military services in line with Truman's view. It wasn't subtle. Johnson had already terminated the building of the Navy's new super carrier, *United States*, which had been authorized by Forrestal. He now mandated a cutback in the overall Department of Defense expenditures for the fiscal year that was about to begin, and notified the services that the Fiscal Year 1951 budget would also be significantly reduced. The Navy, for example, was to have only four attack aircraft carriers in service.

Johnson's directives precipitated what became known as the Revolt of the Admirals. Both Congress and the President acted to put down the so-called revolt in late 1949. On 18 November, Johnson impounded part of the Navy's appropriation for FY1950. The line item for Construction of Aircraft was subsequently reduced from $682 million to $557 million. The impact was exacerbated by the fact that the fiscal year was already 4 ½ months over and counting, which meant that an 18% reduction of the annual budget required a cut of almost twice that in the second half of the year.

As a result, the Navy had to begin cancelling or cutting back existing programs. Because of the recognized need to replace its arguably obsolete propeller-driven fighters with jets as

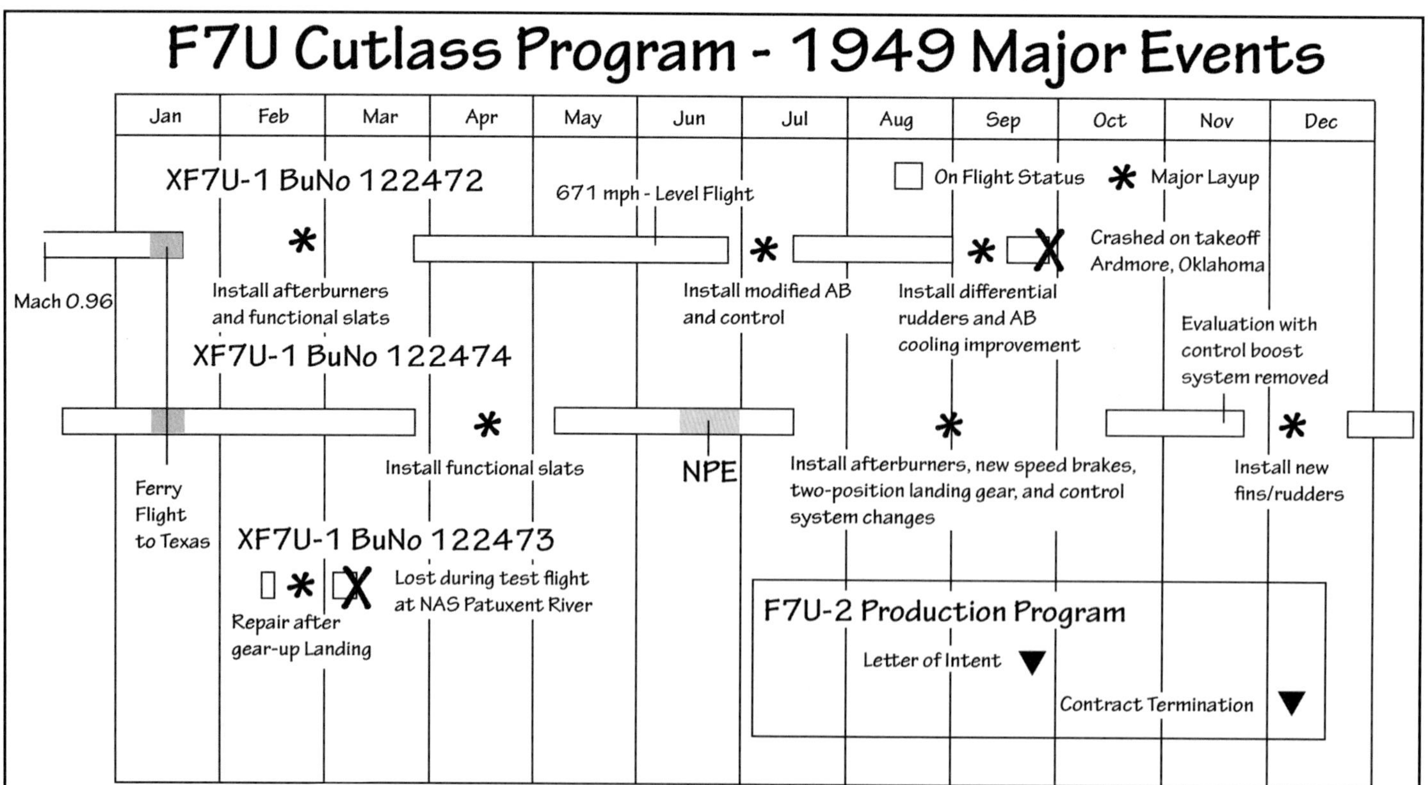

Above, the high-risk, variable-wing XF10F-1, still in development in 1950 as the Navy's general purpose fighter, did not fly until 1952. (Grumman) Below, the F7U-3 became the backup general purpose fighter for the XF10F-1. The first F7U-3, BuNo 128451, and an additional 15 aircraft were built without the benefit of the results of the F7U-1 carrier suitability tests and were built with the early single-wheel nose gear and the bulb-nosed radome. (Chance Vought via Craig Kaston)

quickly as possible, maintaining the production rate of Panthers and Banshees was a high priority. The Navy's other fighters in development—the Jaguar, Skyray, and Demon—addressed critical fighter mission requirements that could not be accomplished by the Panther, Banshee, or Cutlass and so had priority as well. (They also hadn't flown, which meant that their shortcomings and shortfalls were not yet apparent whereas those of the Cutlass were known all the way up to the CNO's office.)

Of all the Navy's jet fighter programs, the Cutlass therefore happened to be the farthest from a chair when the music stopped. The second XF7U had literally disappeared in March, the June NPE had not gone well, and the first XF7U had cartwheeled to destruction in September. The airplane was enough overweight to necessitate a costly redesign that would delay production. Handling qualities issues had not yet been resolved. Developing a satisfactory afterburner had proved to be more difficult than expected. The first production F7U-1 had not yet flown. The CNO had made it clear that he was aware of the existing deficiencies and troubled program status. Given all that, it's not surprising that BuAer terminated the F7U-2/3 production contract on 8 December 1949. The production contract for 19 F7U-1s was not affected since it was funded by the prior year's appropriation.

On 11 December 1949, Vought submitted unsolicited proposals to BuAer for the V-366D powered by two J46s and V-366E powered by one J40–WE-10 engine. According to an internal BuAer document, "Vought considered V-366E to be the better basic airplane design." As it turned out, a J40-powered Cutlass would have been even more ill-starred than the one the Navy eventually opted for, the J46-powered F7U-3.

The week prior to the official notification, Admiral Pride informally advised Vought that the -2s would be cancelled but he expected production to be reinstated with fiscal 1951 funds. However, in January 1950, he announced that the F7U-2 production was "out the window." Vought continued to press for an F7U-3 production order and subsequently provided the Navy with an updated F7U-3 development plan in early February. It was relatively fast-paced, with the first flight of a production airplane only 21 months after go-ahead, because of its structural similarity to the F7U-2, which was virtually identical to the -1. Following discussions with BuAer, Vought submitted a revised F7U-3 proposal at the end of March and began constructing a mockup and writing a detail specification on company funds.

At that point, the Navy's first and only general-purpose fighter, the J40-powered F10F, was well behind schedule. Grumman had only just begun detail design the previous October. It was a fairly high risk design with its untried variable-sweep wing. Vought was proposing a lower risk alternative—both aerodynamically and with the J46, from a powerplant standpoint—that would be deployed sooner that the F10F. So instead of reinstating the F7U-2 production—the only chance it had at fielding a high-performance day fighter in the near future— the Navy elected instead to buy a second general-purpose fighter, the F7U-3. As Vought had suggested, the production F7U-1s, the first of which had just flown on 1 March, were relegated to an F7U-3 design support role, in effect being considered the prototype for the very similar -3 and justifying an early commitment to fairly high rate production.

Vought received an F7U-1 Contract Change Request in April 1950, calling for incorporation of the proposed F7U-3 modifications in the 15th F7U-1 production airplane. This allowed Vought to begin charging the

Above, the first production F7U-1 was assigned to carrier suitability, including at-sea trials. It was a very important and high-visibility milestone for the Cutlass program. Here, XF7U-1 122472 is being removed from Vought's arresting and catapulting static test fixture where it was proof-loaded to qualify the structure for carrier operations. (Vought Heritage Center) Below, the first production F7U-1, BuNo 124415, was completed before parts were available for either the new vertical fin/rudder or speed brakes. However, engineering and parts were created in time for the production tailhook and the afterburner cooling scoops although afterburning engines were not yet available. Ten flights were made before the airplane was laid-up for the incorporation of the production vertical fin and rudder and most of the other changes that had resulted from the flight test of the prototypes. (Vought Heritage Center)

Navy for F7U-3 engineering design and analysis. However, because Vought had accomplished some initial work on company funds, the Navy was able to review the F7U-3 fuselage mockup on 18-21 April. Satisfactory completion of this milestone and the contract change provided Vought with authorization and funding for adding more engineering manpower to the F7U-3 program. The Detail Specification was signed at the end of July.

BuAer provided a letter of intent to Vought on 27 August 1950 for 28 production F7U-3s. It probably wasn't a coincidence that only a few months after the first swept-wing MiGs appeared over Korea in November 1950, BuAer provided Vought with a contract for an additional 212 airplanes, 12 of which were to configured for photoreconnaissance.

Now Vought had to keep the F7U-3 program sold. Continued Navy funding depended on the F7U-1 demonstrating that the problems and issues identified in the Navy flight evaluations and Vought flight test were solved and to a lesser extent, the lack of progress in the Navy's other new fighter programs.

Early 1950 F7U-1 Flight Test

At the beginning of the year, XF7U-1 122474 was the only Cutlass flying. The fin/rudder change was successful, with the pilots noting improved directional "stiffness" and a decrease in adverse yaw with roll; the longitudinal control system change was not. As a result, a new feel system, which introduced the loads that the pilot felt at the hydraulic control actuator valve, was incorporated along with a yaw rate damper in a second month-long layup from mid-January to mid-February. The yaw damper moved a control surface on the right fin below the rudder. In addition, Vought modified the lateral control system to provide higher ailavator deflection to improve low-speed roll response. These changes proved to provide the desired improvement in handling qualities.

The first production F7U-1, BuNo 124415, finally flew on 1 March 1950. (The first production F-86 had flown in May 1948, so the F7U program had slipped another 10 months relative to its Air Force counterpart.) After only 10 flights, it was laid-up for all of April and May for installation of the new fin and rudder configuration and the control system modifications that had been established as a result of the NPE and flight test of the last XF7U-1, as well as installation of instrumentation for the Navy's Carrier Suitability evaluation.

NATC pilots LtCol Marion Carl and LCDR Joe Rees accomplished a

"brief qualitative evaluation" of XF7U-1 BuNo 122474 in early March 1950 at Ardmore. They made seven flights for a total of eight hours flight time. One climb was made with full combat power (afterburners on). The time to 30,000 feet from a standing start was five minutes, which also happened to be the time limit for combat power. It isn't known whether the takeoff was with full internal fuel, 996 gallons, or so-called combat fuel of 633 gallons, which would make a significant difference in both the rate of climb and the endurance after reaching altitude. What was recorded was the fuel consumed in the climb, 300 gallons, which was consistent with the predicted Specific Fuel Consumption (SFC) of two J34s in afterburner.

Above, the first production F7U-1 BuNo 124415 made a second "first" flight in early June after a two-month layup to install the production fins/rudders and control system modifications including the installation of a yaw damper. The incorporation of the speed brake changes was postponed until after at-sea carrier suitablity tests were completed, as was the installation of afterburning engines. Only three weeks later, it was ferried to Patuxuent River to begin the all-important carrier suitability testing. (Vought Heritage Center) Below, Vought test pilot W.P. Thayer (and future president) poses with the first production F7U-1 BuNo 124415 at a photography session following completion. The aircraft was originaly finished with the XF7U-1 short vertical tails but does have the case/clip chutes. (Jay Miller collection)

According to Carl's biography written with Barrett Tillman, "I flew the Cutlass four times and Joe flew it three, and the more we flew it the less we liked it. Consequently, we returned to NATC and recommended that the airplane not be purchased. That started quite a feud." Carl and Rees were critical about handling qualities but agreed that some problems identified during the NPE— longitudinal control sensitivity, low lateral control power at low speeds, and lack of adequate directional stability—had been satisfactorily addressed. Nevertheless, he thought that "the F7U's flight-control system caused problems and its range and carrier-landing qualities left a lot to be desired." He took the trouble to go to Marine Corps Headquarters and voice his views, which resulted in the Commandant approving the position that the Marines "had no use for an airplane that lacked external stores and which couldn't be used in the attack role."

Fortunately for Vought, CDRs Noel Gayler and Charles Gibson of BuAer also flew two flights each and were impressed. Gibson in particular was reported to be quite pleased with the performance and handling qualities, both at Mach 0.91 and at landing speed, saying that he would feel comfortable about taking it aboard a carri-

er after another familiarization flight. Gayler was equally enthusiastic about it.

In a 16 March letter to Captain Trapnell at NATC, Paul Baker made much of the favorable comments from a debrief of the Navy pilots: "The rate of roll was considered good (in the landing condition as well as the clean condition) and the adverse yaw was noticeable but acceptable. The airplane was considered to have good lateral-directional stability and control at all speeds, and lateral dynamic stability was considered satisfactory even without the yaw damper. The pilots commented that with the yaw damper on, the airplane maintained heading in rough air superior to any other airplane flown by them...The pilots considered the rudder effectiveness good and all the ground handling characteristics excellent. Visibility in the carrier approach condition was considered acceptable. The Bureau pilots, CDRs Gayler and Gibson, seemed quite impressed by the low speed handling characteristics of the airplane."

However, the modification to the wing speed brakes and the addition of the fuselage speed brakes had not yet been incorporated, although their design was complete and parts now available. A third month-long layup was therefore accomplished between mid-March and early April to install the fuselage-speed brakes and the modified wing-mounted speed brakes.

In addition to the planned speed brake changes, correction of five "Unacceptable" handling quality concerns was made mandatory prior to the airplane being returned to Patuxent River for Navy flight test:

- Elevator force gradient with speed slopes in the wrong direction

- Longitudinal force instability exists above IMN (Indicated Mach Number) of 0.88

- Neutral static longitudinal stability exists within +/- five to ten knots of trim speed tested except PA (Power Approach)

- Longitudinal force and position stability is too low in NRP (Normal Rated Power) climb and NRP V_{MAX} level flight at an aft center-of-gravity position

- Longitudinal oscillations occur above IMN 0.90 and cannot be damped

Vought expected to be able to correct the first two items by reducing the preload in the feel system and adjusting the ram air input to the feel system with Mach number, beginning at about Mach 0.8. At Mach .95, the input would have been reduced to 25%, reducing the stick force per g from about 40 lbs to 20. In addition, one set of ailavators was to be modified to a blunt trailing edge (7° surface taper instead of 12) and a new seal on the leading edge of the ailavator was to be incorporated and evaluated.

Vought test pilots and engineering did not think the flat spot in static stability worthy of much attention but planned an evaluation to determine if it could be eliminated to the Navy pilots' satisfaction. Otherwise, according to a Vought engineer, "it is...necessary to convince the Bureau that the airplane is satisfactory as is."

The stability problem at aft cg was to be addressed by restricting the aft cg and adding ballast as required.

The pitch oscillation above Mach 0.89 was new. The Vought test pilots had discovered it just before the March 1950 Navy evaluation flights. Engineering suspected that it was caused by minor configuration changes to the engine and afterburner cooling scoops and vents that had recently been made. During flight test the previous year, closing off the forward pair of afterburner cooling ports on the upper fuselage had eliminated a similar longitudinal oscillation that occurred at speeds between Mach 0.77 to 0.80.

Informal evaluation flights in XF7U-1 BuNo 122474 and the second production F7U-1, BuNo 124416 (which had only just flown for the first time), were also made in April 1950 by Navy pilots, including CDR Gayler, CDR Turner Caldwell, CDR Funk, Capt Ramsey, and CDR Cook, who were visiting Vought for the mockup review of the F7U-3.

During one of Gayler's flights, he happened upon a B-36 at 24,000 feet on a production test flight from Carswell AFB and by radio contact with the B-36 pilot, coordinated a brief tactical engagement. Although this F7U did not have an afterburner, he was able to make gun runs on the maneuvering B-36. In his opinion, according to a Vought memo, "two F7U-1s would give the B-36 a fairly rough time by making simultaneous attacks."

Following the Navy evaluation at NAS Dallas, F7U-1 416 was ferried back to Ardmore by Boone Guyton, who had been the Vought project pilot on the XF4U Corsair. While short in duration and limited to 8,000 feet and Mach 0.73 by an overcast, the flight filled Guyton with enthusiasm for the airplane's prospects. "It is recommended that we emphatically point up the ease of flying, excellent stall, low speed approach and landing characteristics, and good ground handling qualities of this airplane."

He went on to note that because of the F4U, "Vought has been considerded a company which can provide an excellent high performance airplane but one which is generally more difficult to fly at or near stalling speeds, during takeoffs, during approaches to and carrier landings, and during normal field landings. Our accident rate has been high. While we eventually erased the poor handling qualities of the F4U airplane during actual carrier landings, spin-stall and landing accidents have continued to prove embarrassing." Ironically, the F7U was to earn an even worse reputation than the F4U's for carrier suitability.

The good news was that the F7U-1 did not exhibit buffet, wing drop, or lateral-directional oscillation at transonic speeds that was typical of other jet fighters at the time. Although the usual trim change was experienced, it was relatively minor: a nose down change became apparent at 0.86 to 0.87 Mach number, reaching a maximum at 0.96 Mach number, at which point it began to diminish. However, at all Mach numbers, stick fixed and free maneuvering stability was always positive with an increase in elevator angle and stick force required to pull g at a constant Mach number. The stick forces, elevator angles, and hinge moments up to Mach 1 were well within the capabilities of the pilot and the airplane. The variation in trim change with speed was now smooth and the maximum pull and push required between onset and Mach 1.0 was only 20 pounds.

The other good news was that the fuselage speed brakes that had been incorporated on 474 were effective, particularly in combination with the wing speed brakes. After opening both at 16,000 feet and 0.80 Mach, the test pilot reported, "it was like running into a brick wall." Only slight buffet and a mild nose-up trim change (which was desirable following the loss of hydraulic pressure to the flight controls) were experienced. With the wing brakes open, an extended dive at a 50-degree angle was accomplished at a constant Mach number of 0.62.

On the other hand, the new longitudinal oscillation proved too difficult to solve, requiring Vought to spend a significant amount of time in April and May 1950 modifying and flying the only two Cutlasses available, the last XF7U-1, BuNo 122474, and the second production F7U-1, BuNo 124416. It was considered "mild", about 1.3 to 1.4 cycles per second (cps) experienced above Mach .92, with a variation in normal acceleration of +/- 0.2 to 0.4 g. While not dangerous, it was a serious shortcoming from the standpoint of the pilot being able to track a target and accurately fire the 20mm cannon.

In-flight tuft studies were accomplished first with 474, using the camera windows in the canopy and cameras in a fairing on the tip of the vertical fins. These did not reveal any airflow disturbance that would cause the oscillation but the engineers suspected that airflow fluctuations over the fuselage that weren't showing up as disturbances of the tufts had to be the cause.

Changes to the fuselage and engine installation were then evaluated on 474 and/or 416. These included:

I. Spoilers of various heights and circumferential locations at fuselage station 300

II. Vortex generators at the top and bottom of the fuselage at station 300

III. Scab-on modification of the aft fuselage to provide essentially flat top and bottom lines on the fuselage and a flat boat tail

IV. Closing the inlet boundary layer duct

V. Several different configurations of afterburner compartment cooling

VI. Reduced engine rpm dives

VII. Vortex generators on the wing just forward of the ailavator hinge line

VIII. Blunt trailing edge ailavators

Static pressure measurement on a small portion of the fuselage did not indicate any fluctuation that would account for the oscillation. It was noted that damping of the short period oscillations deteriorated at higher Mach numbers and was lower than estimated.

Only one of the spoiler configurations and reduced engine rpm was determined to have any appreciable effect on the longitudinal oscillations in 416. However, the spoiler resulted

Below, the frontal area increase when all six speed brakes were extended was significant. Test pilots considered the F7U-1 to have the most effective speed brakes of any of the jet fighters of the time. (Vought Heritage Center)

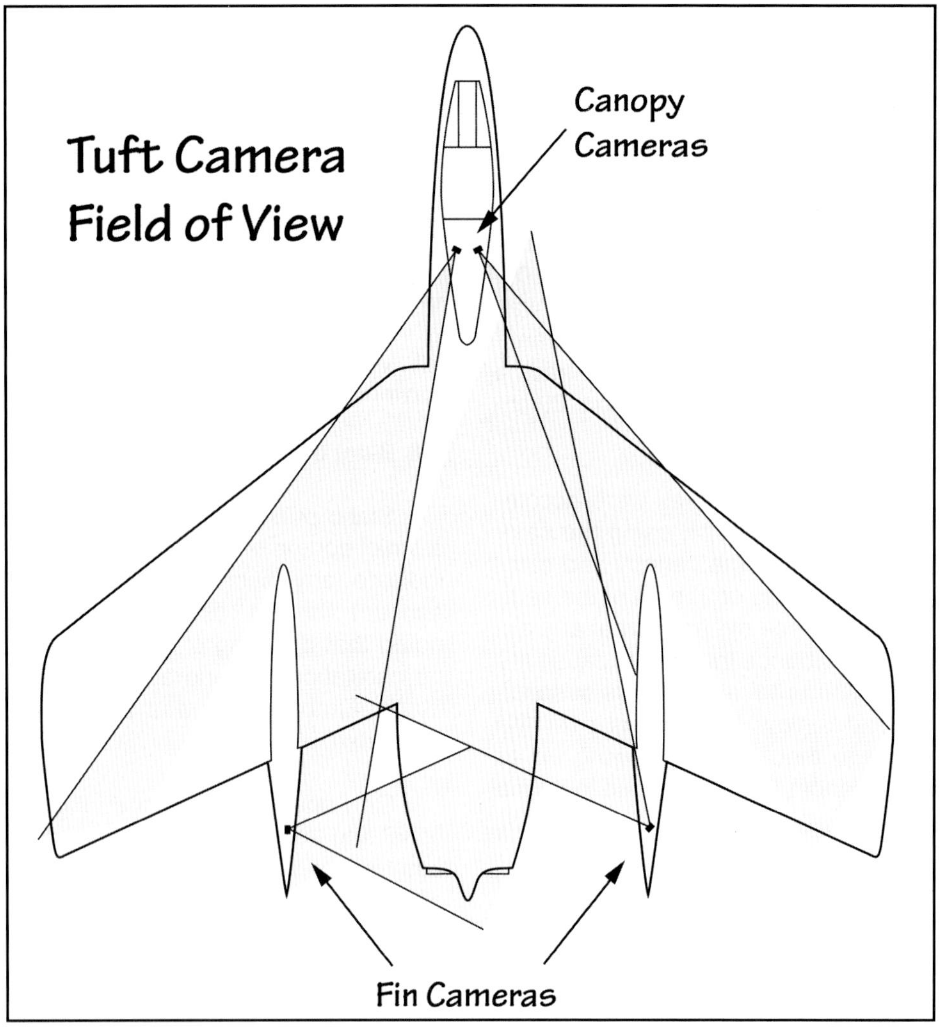

Above, provisions for camera mounts were provided in the canopy and on the inboard sides of the vertical fins to provide images of tuft movement when required for flow visualization. (Author) Below, in late 1941, Westinghouse received a contract from BuAer for a design study of jet engines. Westinghouse had no knowledge of them, but neither did any other U. S company. The result was the Yankee 19A, which first flew under an FG-1 Corsair in January 1944 at Patuxent River. (Jim Sullivan collection)

in an unacceptable increase in drag and 474's oscillations were not diminished by the reduced rpm.

The blunt trailing edge ailavators (the thickness remained the same at the hinge line but the taper angle was changed from 11° to 7° degrees with no increase in the chord of the ailavator) did not have an impact on the oscillations but they were eventually incorporated to increase control power, since they had the aerodynamic effect of increasing the chord of the control surface and therefore resulted in less elevator deflection at high Mach numbers.

In the end, no aerodynamic cause was ever determined for the oscillation. The fix proved to be a pitch rate damper. This was a rate gyro that detected an uncommanded pitching motion and moved the ailavators to oppose it. It was successfully evaluated on BuNo 124416 in June 1950 and victory declared.

In mid-May, the F7Us at Ardmore, 474 and 416, were ferried to NAS Dallas, now the base for Vought flight test operations since the runway extension was complete. The next major phase in flight test was breaking the sound barrier, Mach 1.0. However, Westinghouse was increasingly being an impediment to F7U flight test progress.

You Can't Tell the Engines Without a Scorecard

The J34 was an axial-flow jet engine with an 11-stage compressor, a double annular combustion chamber, and a two-stage turbine. It was used in a number of different airplanes. Dash numbers were used to indicate specific engine configurations—generally intended for a particular airplane—and specified thrust/SFC performance. Like the basic engine designation, "Navy" engines were assigned even dash numbers and "Air Force" engines, odd ones. For example, one of the earliest J34s was the WE-7, four of which powered the Curtiss Wright F-87 night fighter. The WE-19B that powered the Air Force XF-88B was virtually identical to the WE-34 used in the F2H-2 Banshee but modified externally to eliminate local interferences in the engine bay of the McDonnell Voodoo.

A higher dash number didn't necessarily indicate a more powerful engine of that type. It was sometimes necessary to issue a specification for a derated engine in order for it to pass qualification testing so that flight-testing could begin. Since the Navy and

Air Force numbers were assigned in sequence from even and odd lists, respectively, an Air Force engine might have a lower number than an earlier Navy engine, as in the case of the XF-88B's -19.

The original Westinghouse 24C power plant for the F7U was to be designated the J34-WE-22, to which a government-furnished afterburner was to be added. According to Westinghouse Specification WAGT-24C4B-102C dated 7 April 1947, the uninstalled military thrust of the engine, without the afterburner in place, was to be 3,000 lbs; uninstalled normal thrust was 2,290 lbs. .

The production F6U was powered by the Westinghouse J34-WE-30A engine, to which was added a Solar A-103A afterburner. The maximum thrust with the afterburner operating was specified to be 4,100 lbs. Military was to be 3,150 lbs and Normal, or maximum continuous, was 2,640 lbs. Afterburner light-offs were unreliable and at least one takeoff accident was attributed to afterburner failure.

The F2H-2 Banshee and F3D-1 Skyknight were powered by the non-afterburning J34-WE-34, which produced 3,250 pounds of thrust for takeoff and 2,650 lbs continuous. The heavier Skyknight needed more thrust, however, which was eventually provided by the J34-WE-36 engine that, by virtue of being operated at higher temperatures, provided a modest thrust increase of 150 lbs for takeoff and a somewhat more significant 350 lbs increase for continuous operation.

The Solar afterburner performance in the XF7U-1 was disappointing. In late 1948, before the afterburners had been installed and tested, the combat ceiling (500 fpm rate of climb) was projected to be 49,000 feet at combat weight (60% of internal fuel). The afterburners, however, could not be kept operating much above 37,000 feet (they "blew out") and did not provide the predicted level of thrust augmentation. Light-off was unreliable and not even possible above 26,000 feet.

As a result, BuAer decided to have Westinghouse be responsible for both the engine and the afterburner for the production F7U-1, resulting in the J34-WE-32 configuration. It was to have 3,370 lbs of thrust at the military rating and 4,900 lbs of thrust in afterburner, an increase of 740 lbs compared to the original Solar afterburner. The afterburner nozzle was to be electronically controlled and vary continuously in area with thrust rather than simply be fully closed or open.

Above, Westinghouse developed a more powerful version of its Yankee engine, the 19B (19 was the diameter of the compressor inlet). Two were used to power the Navy's first carrier-capable jet fighter, the McDonnell FH-1 Phantom. (NA 80-G-398696) Below, Engine changes were frequent in the early days. Here a J34-WE-22 is being prepared for installation in the first XF7U-1 on 9 November 1948 at NATC Patuxent River. (NA 80-G-397005)

Vought was expecting to receive the

Above, Westinghouse scaled up the 19B to create the 24C for more thrust. It first powered McDonnell's second fighter, the F2D Banshee. Its military designation was J34. (Tommy Thomason collection) Below, early on, the afterburner was simply a separate, bolt-on accessory for the jet engine, replacing the tailpipe that was usually airframe specific. (Author from Vought Historical Center)

first -32s in late 1949 for installation in the first production F7U-1, BuNo 124415.

Unfortunately, Westinghouse had significant problems with qualifying the -32 engine and its electronic controls. Its first test engine suffered a failure in December 1949. Its first attempt at the 50-hr flight qualification test (a prerequisite for flight test) failed in February 1950; another was completed in April but BuAer ruled that the condition of some components examined during the post-test inspection was unacceptable. In part as a result of the impact of the Westinghouse engine qualification problem on Vought's schedule, the Navy decided that 14 F7U-1s would be adequate for test purposes and on 6 April 1950, cancelled the last five that were on order.

Due to the uncertain availability of the -32 engine for production, Vought compiled a matrix of seven different Westinghouse J34 engine alternatives, ranging from the -34 with a tailpipe (no afterburner) at 3,185 lbs of thrust to the -32 with 4,900 lbs of thrust in afterburner. Both Solar and McDonnell afterburners were considered (McDonnell had developed one because available ones were too long for its aircraft) as well as Solar controls on the Westinghouse afterburner. However, each alternative was less satisfactory from some standpoint than the -34 (best availability) and the -32 (best performance).

BuAer therefore authorized Vought to install non-afterburning J34-WE-34 F2H Banshee engines in the first three production F7U-1s, since only the fourth, for power plant test, really needed the -32 engines. This work around was subsequently extended to the fifth and sixth production airplanes. In addition to not having the extra thrust of afterburning, the -34 military rating was 185 lbs less, or about five percent. The fourth F7U-1, BuNo 124418, was completed in early 1950. Vought finally received "highly restricted" X engines in June and eventually installed them in 418 for ground runs that were completed in September. It then sat, waiting for flightworthy -32 engines.

In December 1950, BuAer promised Vought that it would receive three YJ34-WE-32 engines in January and five in February. However, as of early March 1951, only two engines had been delivered to Vought, one on 5 February and the other on 28 February. What was worse, Vought spent almost two weeks of test stand runs of the first -32 engine for a total of eight hours of engine runs including 30 minutes in afterburner of the two hours allowed before overhaul. Several problems were encountered and corrected before Vought was finally satisfied that the "engine appears acceptable for limited experimental flight pending satisfactory ground runs in (BuNo 124418)" although it was noted that maximum afterburner thrust appeared to be approximately 100 lbs below specification. Only one -32 would be installed in 418 initially. Its other engine would be the non-afterburning -34. 418 finally flew for the first time in late March 1951.

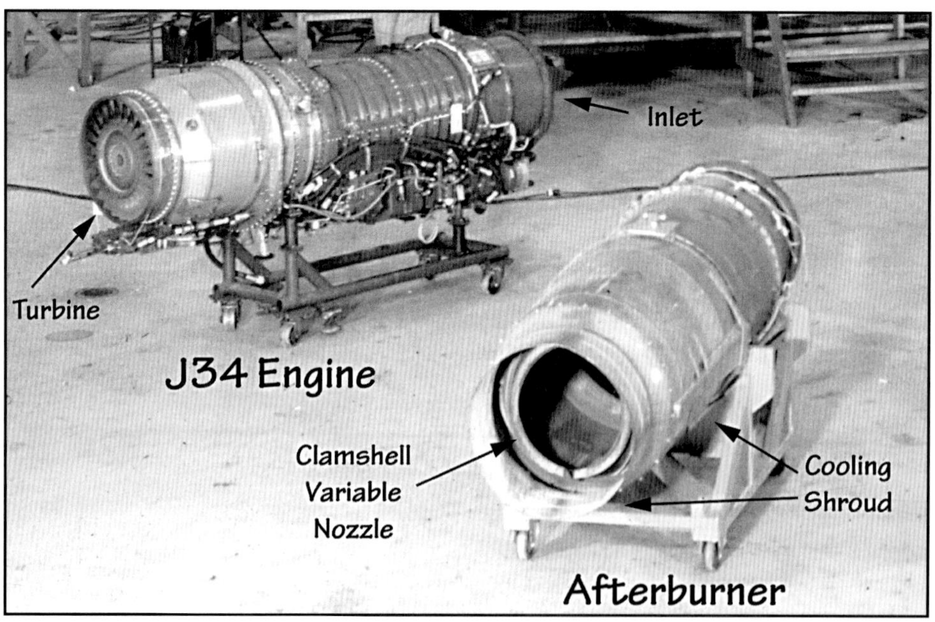

Exasperated, Vought wrote BuAer the following in a 19 June 1951 letter:

"The contractor has been attempting to prepare F7U-1 airplanes for delivery for the past several months under the handicap of unsatisfactory power plants. The J34-WE-32 engines have been well over a year late and have not achieved any reasonable degree of reliability. Engine deficiencies called to the attention of the engine manufacturer have not been fixed, nor have promises for fixes been forthcoming.

The delay in the power plants has not only delayed airplane deliveries,

but has made it impossible for the contractor to follow production and flight test plans. The delay has been costly and now promised to be beyond the contractor's ability to support financially. In addition, the delayed F7U-1 work is conflicting with the contractor's F7U-3 project.

Accordingly, the Contractor wishes to advise the Bureau that no F7U-1 flight operations with J34-WE-32 engines will be conducted after 30 June until more positive assurance is received that the J34-WE-32 engine is airworthy and suitable for the completion of the flight test program or a satisfactory alternative has been agreed upon.

The Contractor respectfully requests a conference at BuAer as soon as possible to review the remaining F7U-1 program in the light of the power plant situation and to discuss possible alternate power plant arrangements. It is considered desirable to review the F7U-3 power plant situation at the same time."

BuAer responded promptly, inviting Westinghouse and Vought to a conference in Washington on 29 June 1951 to discuss the ongoing problem and agree on a plan to go forward.

At the conference, Vought engineers proposed to combine the -34 Banshee engine with the -32's afterburner and a simplified engine and afterburner control system. Capt. Ramsey, head of BuAer's Piloted Aircraft Section, proposed a detailed aircraft-by-aircraft schedule to accomplish development of the fuel controls in parallel with Vought flight test. Unfortunately, Westinghouse was not prepared to comment on its ability to support Capt Ramsey's plan other than to point out the shortcomings and risk associated with any alternative to their continued development of an electronic fuel control system and infinitely variable afterburner nozzle.

In September 1951, BuAer finally decided to give up on the Westinghouse electronic fuel control and accept no more -32 engines. The interim solution was a -32A configuration that eliminated the electronic control and utilized a hydro-mechanical control that was basically the existing -32 emergency control system. The turbine from the -36 engine that powered the F3D Skyknight was also to be incorporated to improve engine acceleration. Only seven more -32As were to be delivered with the remainder of the F7U engines to be either non-afterburning -34s or what was initially referred to as the -32B that was now to be designated the -42. It would have a Holley R46 hydro-mechanical fuel control for the basic engine and simplified management of afterburner operation. Instead of being infinitely variable, the nozzle would now have only two positions: closed down for non-afterburning operation and fully open for afterburning. Afterburner fuel control was to be accomplished with a flow-sensing device incorporating temperature trim.

Below, Vought's F7U program was plagued by engine development and qualification problems at Westinghouse. The F7U-1s in most pictures do not have afterburners, because there weren't many being delivered to Vought until the flight test program was well under way. The two assigned to the Blue Angels in 1952 did not have afterburners, for example. (Author)

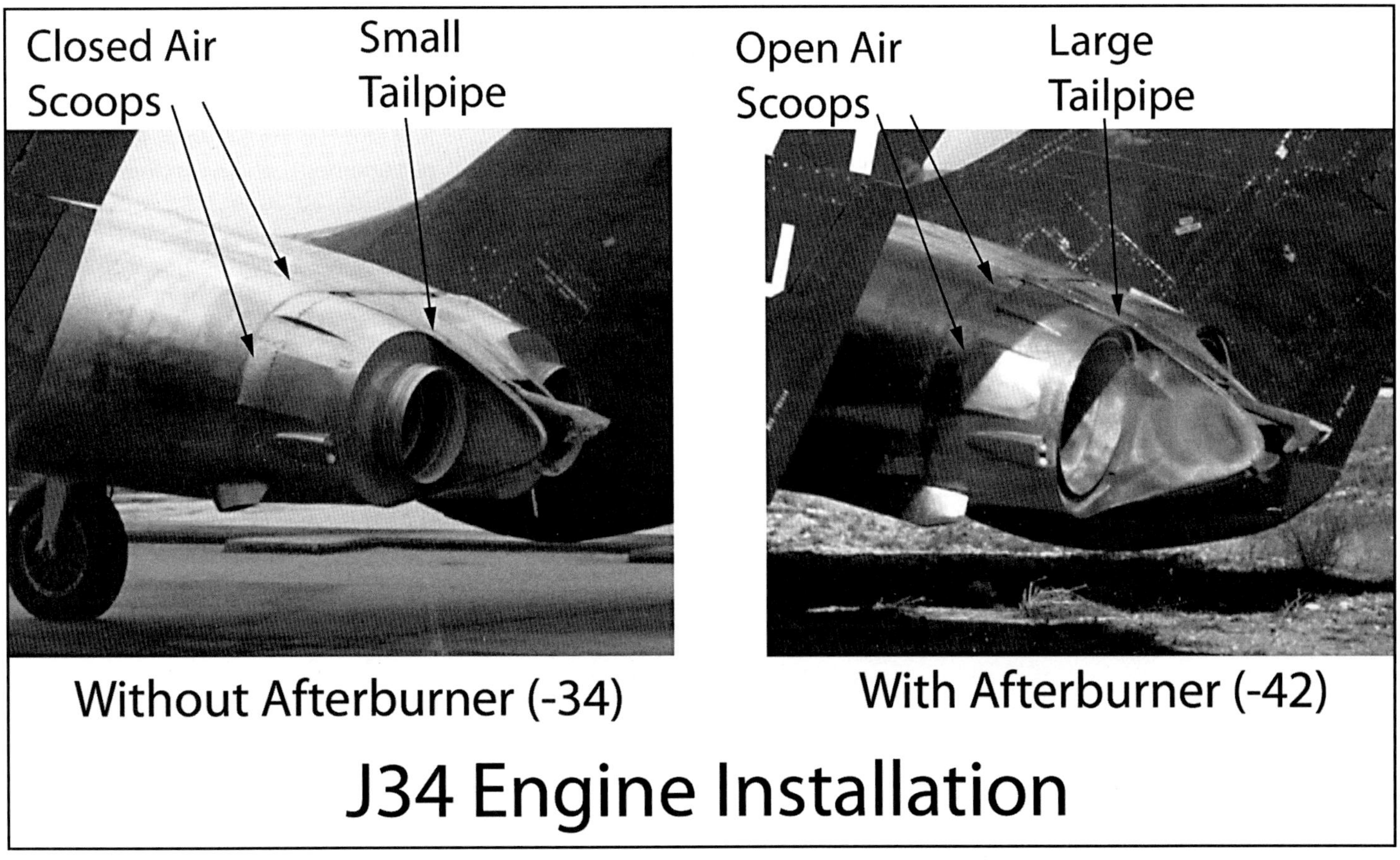

Vought received one prototype -42 in August 1951 and another in October. One of two production engines promised for February 1952 was delivered that month (it suffered minor damage to the afterburner during test runs at Vought) and the other delivered in March in lieu of the two promised for that month. By the end of July 1952, Vought was to have received a total of 26 -42 engines; in fact, only 14 arrived. In the end, most of the surviving F7U-1s were retrofitted or completed with the -42s. When the last F7U-1s at Vought were being prepared for delivery with two -42 engines, the production test pilots routinely made supersonic dives as part of the acceptance flights.

The -42 had to be rated lower than the -32 in order to pass the qualification test. The maximum thrust in afterburner was reduced by 250 lbs to 4,650 lbs and the military and normal thrust by 170 lbs. (The military rating of the -36 engine was 3,400 pounds, but its thrust was not restricted by the afterburner fuel spray bars and flame holders.)

Notwithstanding the trouble it caused Vought, the J34 would prove to be a reliable, durable engine for the Navy, albeit without afterburning. (None of the Air Force jets that were powered by it reached production but in most if not all cases, the engine was not the reason). In addition to long careers powering Banshees and Skyknights, it would be used as an auxiliary engine on the P2V Neptune and to power the single-engine version of the North American T2J carrier-capable trainer. These engines were rated at 3,400 lbs takeoff thrust like the -36.

From High Point to Low Point

XF7U-1 BuNo 122474 had to be used for most of the high Mach number flights in mid-1950 because it was the only F7U configured for afterburners at that point. Its J34-WE-22 engines were removed and replaced with the -34 engines incorporating the existing Solar afterburners. The configuration of its afterburner cooling system was modified. The XF7U-1 originally had 10 small holes in the top of the fuselage with hinged doors that were opened or closed depending on airflow. However, when the afterburners were finally installed on the XF7U-1, a revised cooling arrangement was incorporated. Two of the small holes were sealed shut, four were changed to suck-in doors for ground cooling, and four were changed to open out for pressure relief. Two openings were added in the bottom of the engine compartment along with four air scoops approximately one inch high on the aft end of the fuselage.

As one of the final modifications before the supersonic flight, modified hydraulic pumps were installed that raised the control system pressure to 2,900 psi from the original design pressure of 2,600 psi. (The production airplanes had a boost system pressure of 3,300 psi.)

Even with afterburning, Mach numbers above 0.9 could only be reached in a dive. For example, when Paul Thayer broke the sound barrier in an F7U for the first time on 21 June 1950, he rolled 474 inverted from level flight at 33,000 feet and 0.9 Mach and pulled to a dive angle of about 47° and Mach 0.95, at which point he rolled back upright. Mach 1.006 was reached at an altitude of approximately 26,500 feet, at which point he began a pullout. He pulled to 1.5 g at Mach 0.99 and 2.0 g at Mach 0.975 in the process of returning to level flight at 18,000 feet.

Higher starting altitudes were not possible because of the likelihood of afterburner blowout above 33,000 feet. Mach 1.006 was therefore not considered to be the maximum speed attainable. The combination of maximum starting altitude of 33,000 feet and the pilots' desire to be back in level flight no lower than 18,000 feet made it the stopping point for the time being.

On 24 June 1950, the now supersonic XF7U-1 474 and the first production F7U-1, BuNo 124415, were ferried to Patuxent River, the former for a Navy evaluation of the control system and speed brake changes and the latter to commence carrier suitability trials.

It was the high point of the first six months of 1950. The program was back on track. But not for long.

On 6 July, Vought test pilot Warren Smith was returning to NAS Dallas in the second production F7U-1, BuNo 124416, following an engine failure when he crashed. It was the first flight of the airplane following the replacement of an engine that had failed on 30 June, also requiring him to make a single-engine landing. Tragically, on his approach this time, the F7U nosed over from level flight at about 400 feet above the ground and dove straight into the ground. From the wreckage, the test instrumentation installed, and eye witness accounts, it was determined that the

At left, the first production F7U-1 was ferried to Patuxent River in late June 1950 to begin its carrier-suitability trials. The Navy markings include NATC on the vertical fin for Naval Air Test Center and FT for Flight Test on the nose along with the last three digits of its Bureau Number. (NA 80-G-421729)

slats were extended, the landing gear was up, his center of gravity was farther aft than desirable due to an existing and undetected malfunction in the automatic fuel system transfer timing, and his airspeed was about 175 knots.

The combination of configuration, cg, and airspeed was an unfortunate one, because it meant that there was a significant nose-up trim requirement being provided by the ailavators. At that point, he reduced the throttle on the good engine to idle and began to extend the speed brakes to slow down for landing. Since the hydraulic pump on the failed engine was not providing any flow, the combination of a reduction in engine rpm and increase in flow requirement to the speed brake actuators caused the hydraulic pressure to drop. The loss of pressure was apparently enough to result in the automatic bypass of the control actuator hydraulic fluid so his ailavators were no longer boosted in the trim position necessary to counteract the nose-down pitching moment in that configuration at that speed. It was not enough, however, to trigger the unlocking of the spring tab from the ailavator surface and its connection to the mechanical control system.

Unfortunately, the design of the automatic reversion to manual control assumed that both changes necessary for direct pilot control of the spring tab on the ailavators would occur when hydraulic pressure was lost. Since the ailavators were now unboosted, but the spring tab wasn't providing any help, the pilot would have had to pull back on the stick with 275 lbs of force to move the ailavator back to the required trailing-edge-up angle. At 400 feet above the ground, that wasn't even close to possible before the airplane had pitched too far nose down to recover.

In hindsight, a similar if not identical combination of airplane configuration, speed, and the failure of the control system changeover to occur as intended was now suspected to have been the cause of Millar's fatal accident in 122473 in March 1949.

The very next day, Paul Thayer had to eject from the third (and last remaining) XF7U, BuNo 122474, at Patuxent River. It was, as Navy pilots say, colorful. He was participating in an airshow and after making his second high-speed pass in afterburner and pulling into a vertical "S" maneuver completed at 8,000 feet, he noted that the fire warning light for the right engine was on. The airplane was in fact on fire, possibly due to a leak in an afterburner fuel line. He had shut down the right engine and begun a descent for an emergency landing when the left-hand engine fire warning light came on and it became apparent that it was failing as well. He shut down that engine and was considering a dead-stick landing but after determining that "the emergency aileron boost system was also inoperative" and noting that there were "heat waves and what appeared to be smoke or vapor" emanating from the aft fuselage, he decided that it was time to abandon the airplane. (Unbeknownst to him because his radio had failed, the tower was advising him to eject.) Not trusting the virtually untried Vought ejection seat, he initially tried to simply bail out but was unsuccessful in getting out of the airplane due to g forces. He then ejected without bothering to strap back in and parachuted to safety.

Above, the only picture that I can verify is 12416 is unfortunately not a very good one. It first flew in early April 1950 and crashed three months later on 7 June 1950, the victim of an unforeseen configuration of the F7U-1's flight control system. (Dave Stoddard via Tony Landis) Below, F7U-1 BuNo 124417, shown here in the Vought paint shop, had a very short career. Because it was laid-up for four months shortly following its first flight, being updated to the production configuration, it was only on its ninth flight when Vought test pilot William Harrigan crashed it on landing at NAS Dallas following a hypoxia-created in-flight emergency. (Vought Historical Center)

Two weeks after Smith's accident, Vought had completed its evaluation of the cause and BuAer had agreed to the following set of restric-

tions, procedures, modifications, and maintenance checks:

I. Slats were not to be extended before the landing gear was fully down and neither was to be extended at excessive speed.

II. The trim tabs were to be set at 10 degrees for takeoff and landing.

III. The fuel system transfer timing was to be checked.

IV. The hydraulic tank capacity was to be increased.

V. A pressure-operated valve was to be installed that cut off hydraulic flow to the utility system components when pressure fell to a minimum level for flight control actuation.

VI. Tolerances on the boost cylinder bypass valves were to be tightened.

VII. The changeover spring was to be strengthened.

Aircraft No. 7, BuNo 124421, was pulled from final assembly to be used for engineering test and evaluation of the control system changes.

BuNo 415 was ungrounded on 21 August after boost system modifications were accomplished on it at Patuxent River. For a short while, it was the only F7U-1 that was flyable. In mid-September, BuNo 124417, the third production F7U-1 and the replacement aerodynamic demonstrator for the ill-fated 416, would briefly join it on flight status.

Less than two weeks later, on 28 September 1950, Vought test pilot William Harrigan was evaluating speed-brake operation at high speed and altitude in 417. He had successfully demonstrated a changeover to manual boost at Mach .9 when he reported that he was shutting down the left-hand engine and coming back to NAS Dallas for an emergency landing. Landing to the south, he was fast on touchdown and could not stop before the airplane rolled off the end of the runway into Mountain Creek Lake. Tragically, he had kept the canopy closed for landing and the failure of the nose landing gear strut during the overrun jammed the canopy shut. He drowned before rescuers

Below, given the trials and tribulations of F7U-1 flight test from July to September 1950, only the Navy's urgent need for a swept-wing carrier-based fighter (the Korean War had begun at the end of June and the MiG-15s would arrive in November); some successes (breaking the sound barrier and a boost change-over at Mach 0.9); and the status of its other high-performance fighter programs kept the F7U-3 program alive. (Author)

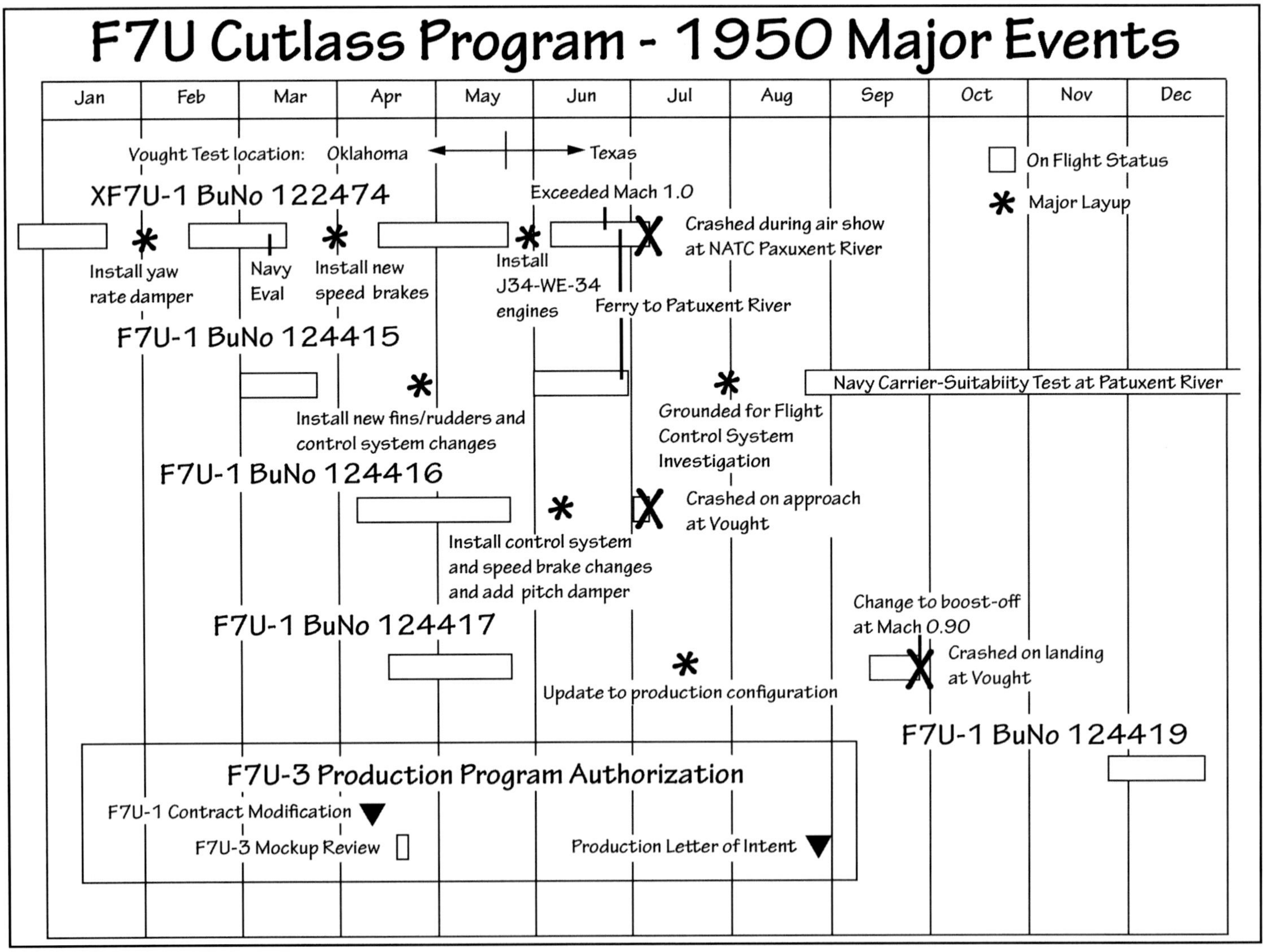

could reach him.

It was subsequently determined from a review of the radio transmissions, evaluation of the instrumentation records, and discovery that the oxygen bottle was empty, that Harrigan was probably suffering from anoxia. In that state, he apparently misunderstood cockpit instruments and lights, shutting down a good engine and making a poor approach and landing. As a result of the crash crew being unable to quickly open his canopy, an emergency external canopy release was subsequently retrofitted to the F7Us.

With all the XF7U-1s gone along with BuNos 416 and 417 and BuNo 418 waiting on production engines, BuNo 415 was once again the only F7U-1 on flight status. It was at Patuxent River for the carrier suitability evaluation and therefore not available for Vought's development work. Vought would finally resume flight test in late November with BuNo 124419 and in January 1951 with 124420.

Fortunately for Vought, the BuAer's other fighter programs weren't fairing any better in late 1950 so the Navy was in no position to waver on its commitment to the F7U-3 production program due to the spate of F7U-1 accidents.

Initial Carrier Suitability Testing Problems

Carrier suitability testing finally commenced with land-based evaluations, both takeoffs and landings, with BuNo 124415 in mid-September 1950. The major problem noted was with the arresting hook assembly. By the end of January 1951, 94 arrested landings had been accomplished and four problems had been identified:

I. Grooving of the chromium-plated hook throat
II. Premature unlocking of the V-brace.
III. Failure of the inboard hook fairing
IV. Failure of the tendon cable

Facing the hook throat with colmonoy rather than chromium plating it increased the durability of the hook throat; this was made feasible because the cable shedder was determined to be unnecessary and deleted.

Above, the F7U-1 was originally to be hooked-up to the catapult shuttle at a single point, which was also where the airplane end of the holdback would be attached. This meant that only one strong point for catapult hook-up had to be provided on the bottom of the fuselage. (NA 80-G-421713) Below, as originally designed, the catapult holdback was attached to a cleat on the deck immediately under the engine exhausts, at the point where clearance with the bottom of the fuselage was least. (NA 80-G-421711)

The premature unlocking of the V-brace was minimized by changing the trail angle of the hook so it didn't push the V-brace upward when it contacted the deck. It sometimes also occurred when the hook point hit the non-flush fiddle-bridge base plate on the shore-based arresting gear installation before engaging a cross-deck pendant. Since the V-brace was not damped, this could result in the hook point skipping over at least a few subsequent pendants, risking a barrier engagement.

The inboard hook fairing was finally removed from the hook shank and hinged on the right-hand V-brace fairing.

The tendon cable failure was termed the greatest single source of annoyance and delay in the shore-based carrier suitability test. Vought finally agreed to redesign the complex hook mechanism to eliminate the tendon cable, which was required to 1) retain the V-brace and hook shank as a unit long enough to drop the cross-deck pendant after arrestment and 2) lock the V-brace in the closed position.

Buffet Boundary

One of the notable compressibility effects, in addition to a significant increase in drag, was buffet of the airframe and the control surfaces at high altitude. The buffet was particularly apparent in a hard turn or pull-up and increased as the level of g increased,

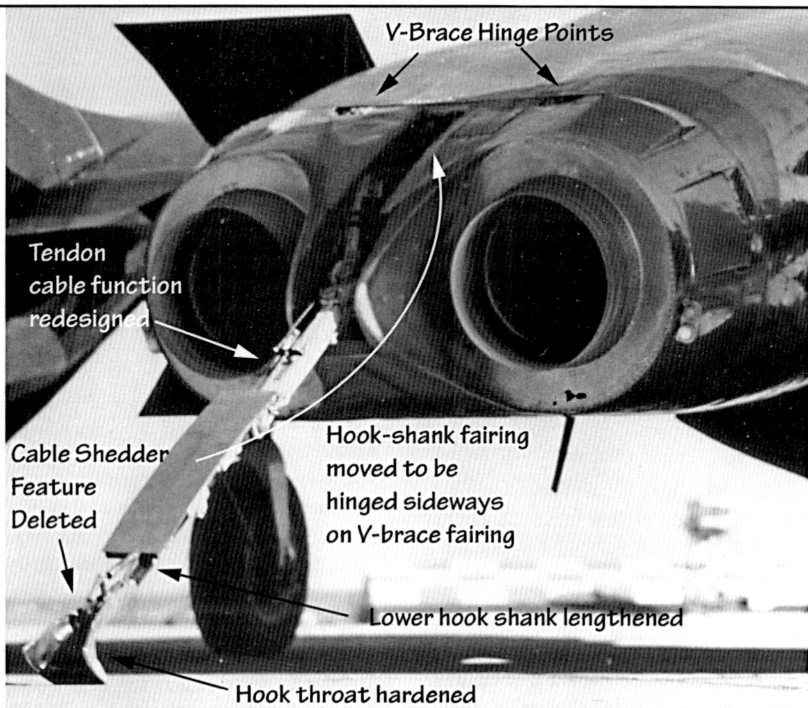

Production F7U-1 Tailhook Modifications

Above, numerous changes were required to the complex production tailhook mechanism as a result of shore-based arrested landing tests. (Author)

becoming so heavy that gun tracking accuracy was unacceptably degraded.

Relative buffet boundary comparisons were generally made at 35,000' and Mach 0.9. The F-86A was used as the baseline by Vought engineering in the early 1950s. The Sabre experienced the onset of buffet at about 2 g; the gun tracking limit buffet was reached at approximately 3 g with maximum g being 3.5. For the F7U-1, onset was at a bit lower g than the F-86 below Mach 0.9 and higher above. However, it was decidedly inferior to the F-86 from the standpoint of limit buffet for gun tracking, about 0.8 g less at Mach 0.95. It was about 0.5 g worse from the standpoint of maximum g capability from Mach 0.8 to 0.95. The difference was attributed to the greater wing thickness-to-chord ratio of the F7U-1.

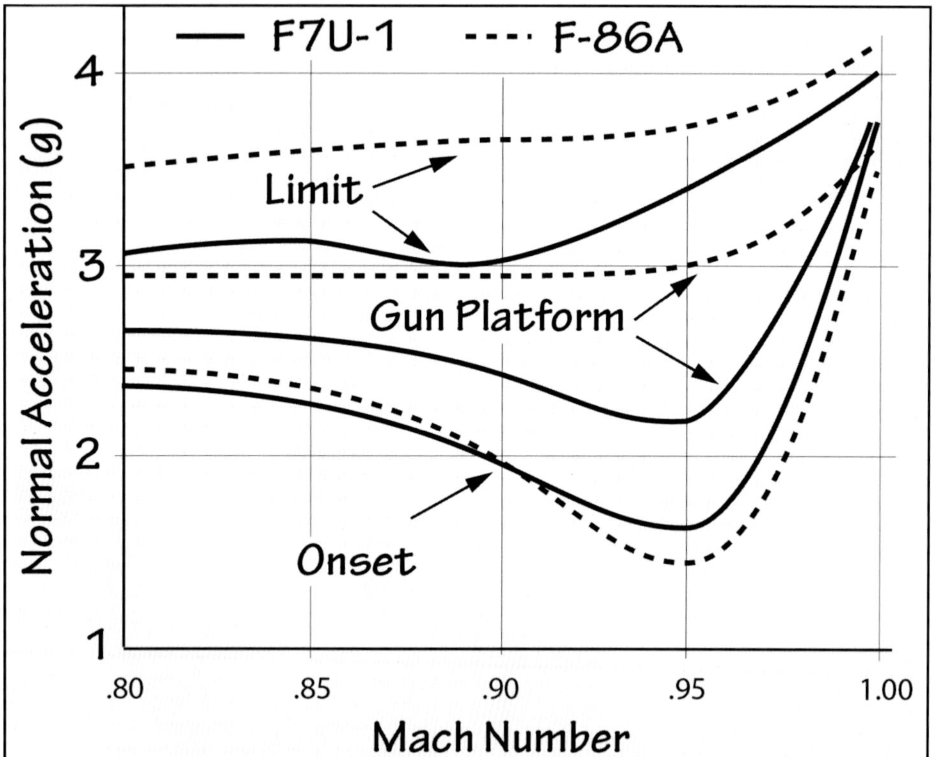

Buffet Boundaries - 35,000 Ft

Various modifications to improve the buffet boundary were evaluated, including addition of a dorsal fin and locking and sealing the leading edge slats. None were successful.

For some reason and indicative of the lack of complete understanding of the transonic aerodynamics, the No. 6 F7U-1, BuNo 124420, had a buffet onset about 1 g higher than the other -1s, albeit within a narrow speed range: it was no better below 0.87 or above 0.92 Mach number. When its outer wing panels were installed on the No. 4 F7U-1, BuNo 124418, its buffet boundary improved by about 0.5 g. If a reason was determined for the improvement provided by those particular outer wing panels, it is not available in the files.

Strong But Not Unbreakable

Flight 28 of BuNo 124420, the structural demonstration airplane, almost ended disastrously in April 1951. The pilot was pulling out of a high Mach number dive at 37,000 feet when the upper part of the left fin and rudder failed and departed the airplane. The flight record indicated rudder buffet at the time of the failure.

The engineers already had a concern about the calculated fin loads being underestimated for the rolling pullout condition so a flight restriction for that maneuver was in place until testing in the Part II structural demonstration confirmed the loads were less than calculated. The failure was therefore doubling concerning.

Buckles on the skin indicated that

the fin that failed had been subjected to just less than the design ultimate load in an inboard direction and the right tail had been loaded inboard to approximately 75% of the design ultimate load. The predicted load was significantly less. Most of the measured load, however, was determined to be oscillatory, i.e. vertical tail shake at high air speeds due to inadequate torsional stiffness, rather than the predicted steady-state loads that had been used to design the structure.

After extensive investigation, including flight test of the existing tail fins between 16 May and 7 June 1951,Vought decided to beef-up the vertical fins either by replacing the Fabrilite sections with aluminum structure or by adding an external strap between the upper and lower sections of the fin. Flight tests of the stiffened and strengthened vertical tails were accomplished in October and November 1951. The results cleared the redesign for the original flight envelopment.

Above, the U.S. Air Force provided a few North American F-86s to NATC for use as chase and pace (airspeed calibration) airplanes. It also represented the state of the art for the maneuver buffet boundary vs. Mach Number. (Author collection)

On 2 November 1951, BuAer updated the "XF7U-1" Fight Restrictions. The maximum "recommended" gross weight for takeoff from smooth, paved runways was 24,000 lbs; the corresponding landing weight was 17,000 lbs. Maximum indicated airspeed limitations were:

Landing gear extended	**175 knots**
Slats extended	**150 knots**
Canopy open	**260 knots**
(no restriction for emergencies)	
Arresting hook down	**150 knots**

"In order to maintain adequate (longitudinal) stability, slats must be retracted before landing gear, and landing gear must be extended before slats."

The maximum g limit was dependent on gross weight, airspeed, and altitude. For 19,800 lbs, it was reduced to 4 at up to 20,000 feet and only 2.5 at 40,000 feet and even then, only at 230 to 240 knots indicated. (The maximum indicated airspeed allowed at 40,000 feet was only 280 knots, but that equated to approximately Mach 0.94.)

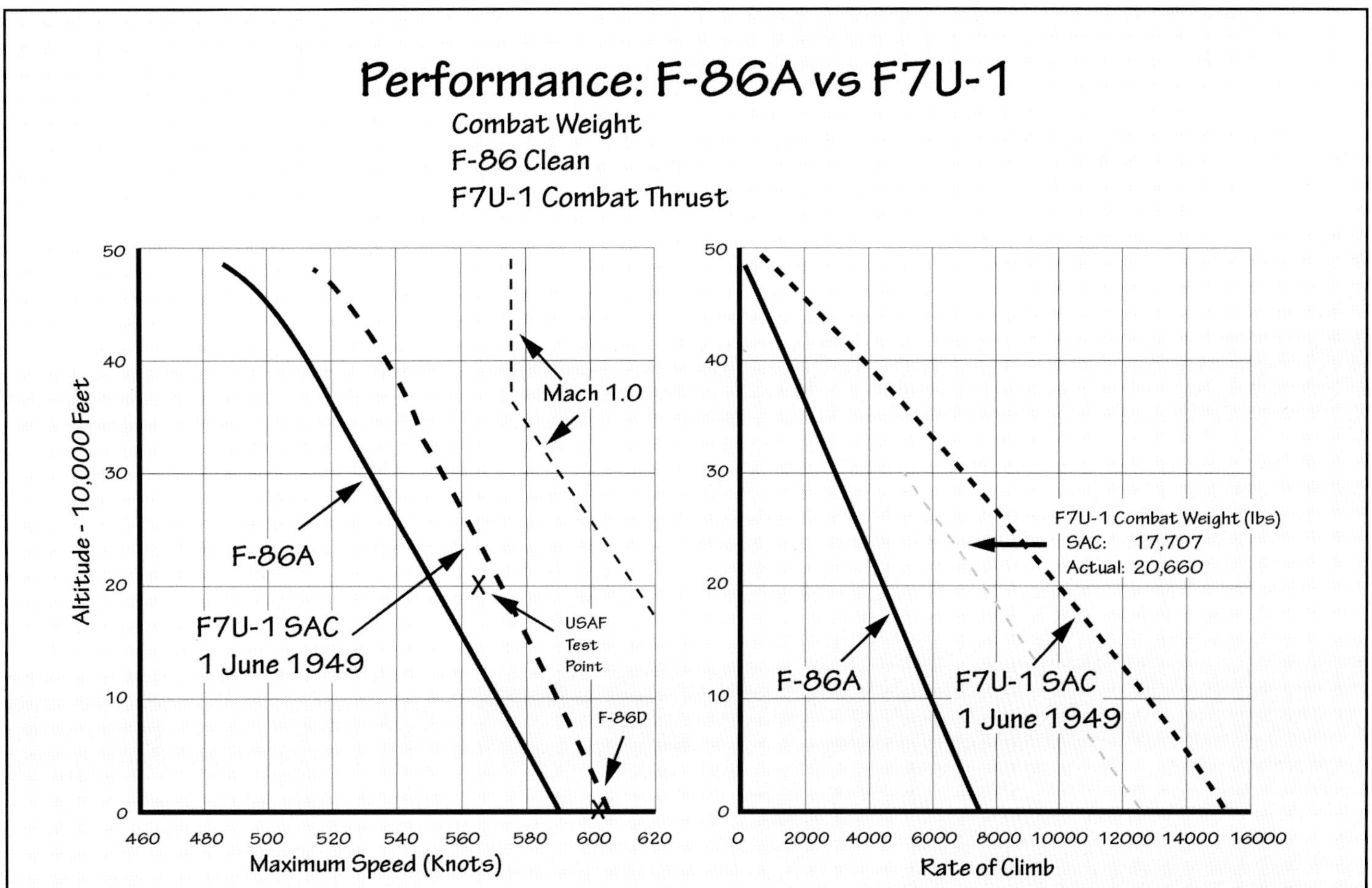

Above, following the inflight failure of its left vertical fin, 124420 was repaired with all-metal upper fins and reflown for a fin-loads evaluation. (Vought Heritage Center) Below, some F7U-1s simply had their vertical fins reinforced with an external strap like this Blue Angel example. Note that the strap on the right-hand fin extends higher than the strap on the left-hand fin because the junction of the upper Fibralite portion of the fin was higher on the right-hand fin for some reason. (Vought Heritage Center) Bottom, F7U-1 BuNo 124421 was initially used for some miscellaneous test work before being modified for F7U-3 catapult concept testing. (Author's collection)

Support of F7U-3 Design and Development

Although the first F7U-3 was to fly at the end of 1951, F7U-1s were to be used to support -3 design and development to expedite its deployment to the fleet. BuAer amended the F7U-1 contract to provide for the modification of two F7U-1s to serve as F7U-3 surrogates. The No. 7 F7U-1, BuNo 124421, was to be used for a full-scale evaluation of the proposed F7U-3 catapult concept; it was to be restored to the F7U-1 configuration following the tests. The No. 9 F7U-1, BuNo 124423, was to be used to test the dual power control system, the rocket pack configuration, and the external stores; it would be delivered in this configuration.

124421 first flew on 24 March 1951. It was first used for new pilot familiarization, a fuel system evaluation, and testing of the Goodyear solid-rubber tire that had first been proposed for the V-362. It was laid up in late June for modification to the catapult test configuration. Different concepts were to be tested that allowed for maximum lift at the end of the catapult stroke, both dynamic rotation during the stroke and rotation to the maximum possible angle for launch prior to the stroke beginning. It was back on flight status for September but a turbine failure delayed its arrival to Patuxent River for F7U-3 configuration catapult tests. Testing at Patuxent was accomplished from November 1951 through July 1952 when it was returned to Vought for conversion back to the standard F7U-1 production configuration.

The so-called aerodynamic prototype, 124423, was first flown on 21 September 1951. From then through early June 1952, it was used for evaluation of various flight control system improvements such as the dual power actuation, a modified feel system (trimmable centering spring, viscous damper, and bob-weight), the F7U-3 speed brake design, an emergency feel system, long chord ailavators, etc. The results of the control system evaluation were first incorporated on F7U-3 No. 5. It doesn't appear that any external stores or the rocket pack, the availability of which was delayed by a redesign to eliminate interference with the F7U-3 catapult bridle arrangement, were ever installed on the airplane.

In addition to those two F7U-1s designated as F7U-3 prototypes, other F7U-1s were also used to support the F7U-3 program. In January 1952, for example, 124418 was flying with control system modifications to improve control feel and vortex generators on the wing to determine their effect on approach speed and stall characteristics, in addition to evaluating the -42 afterburner. Beginning in February 1952, BuNos 124424 and 124425 were assigned to engine and afterburner development, primarily to support Westinghouse's J46 engine program. BuNo 124424 was used to flight test prototype J40 and J46 engine controls using a -32A engine. At one point, the J46 automatic turbine outlet temperature control was installed on a J34 engine modified with a J46 iris-type afterburner and flown in military and afterburner between 30,000 and 38,000 feet.

Navy Service Acceptance Trials

By mid-1951, the Navy had decided to cancel most of the Board of Inspection and Survey (BIS) trials planned for the F7U-1 as well as the assignment of four F7U-1s to VX-3, the operational evaluation squadron based at NAS Atlantic City. Since there were only a small (and dwindling) number of F7U-1s that were to be used exclusively for test, evaluation, and F7U-3 prototyping, the CNO decided to redesignate the production F7U-1s as XF7U-1s. However, the carrier suitability trials were still to be accomplished as well as abbreviated stability and control and aircraft/engine performance trials to determine whether the NOa/s/ 9937 contract guarantees had been met. (The X redesignation proved to have unintended consequences and was rescinded on 17 January 1952.)

The Service Acceptance Trials report on the Model F7U-1 Aircraft

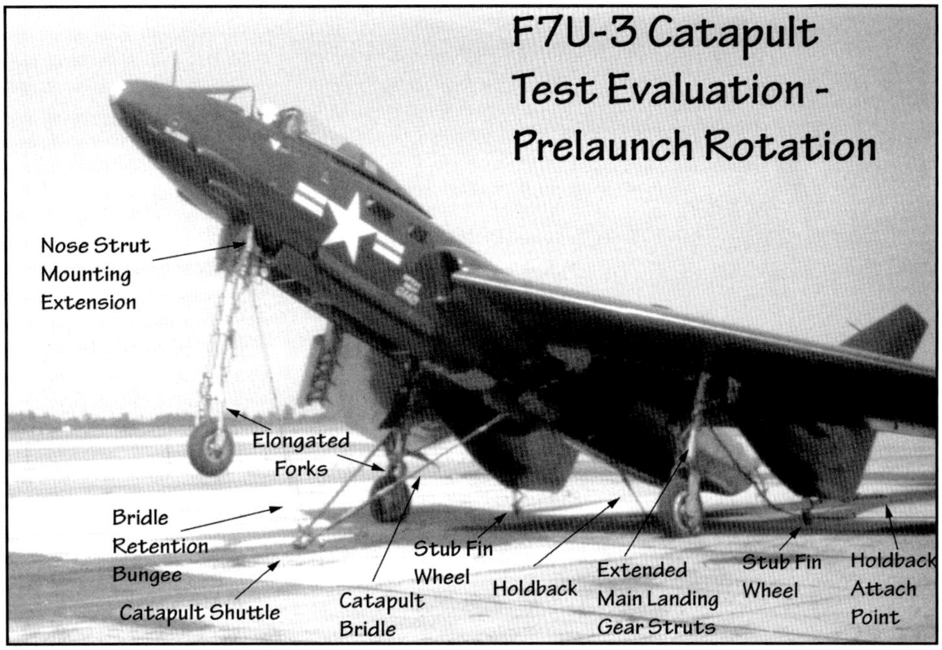

Above, for F7U-3 catapult concept testing, BuNo 124421's landing gear was modified to represent the length of the F7U-3's. Here, BuNo 124421 has been tensioned up with the holdback cable to achieve the maximum nose-up attitude permitted by main landing gear oleo extension and the small tail wheels mounted on the bottom of the stub fins. The bridle-retention bungee line between the catapult bridle and the nose landing gear shortened the time required to hook the Cutlass up for the next test launch. (Author) Below, F7U-1 BuNos 124424 and 425 were used by Vought to evaluate Westinghouse J46 engine development. (William Swisher) Bottom, Vought continued to produce F4Us while developing the F7U-1, in this case a French F4U-7. The canopy of BuNo 124423, which was configured to the F7U-3 flight control system design, appears to have been modified with a radio antenna. (Vought Heritage Center)

U.S. NAVY
418

NAVY

415
422

415 NATC

stated: "The F7U-1 failed to meet contract guarantees for weight empty, performance, carrier suitability, stability and control, and general suitability for service use from the standpoint of maintenance and operational reliability."

Only two F7U-1s, BuNos 124415 and 419, actually wound up being used for the formal service trials. Patuxent River assigned 415 to carrier suitability evaluation and aircraft/engine performance testing; 419, to stability and control tests. 124422 was the designated backup to 415 and was ferried to NATC in July 1951 for that purpose.

Carrier Trials

The basic technique for landing a jet airplane on a carrier was the same as it had been for decades for a propeller-driven airplane, but there were some subtle differences. The final approach was similar to that for a short-field landing, "dragging it in" at minimum altitude and airspeed as coached by the landing signal officer (LSO). When the LSO judged that "cutting" the power would result in the airplane clearing the deck edge, he signaled a "cut" to the pilot, who responded by immediately reducing

At left top-to-bottom: F7U-1 BuNo 124418 was also retained at Vought for miscellaneous development and evaluation efforts in support of the F7U-3 program. (Vought Heritage Center) (Vought Heritage Center) The first F7U-3, BuNo 128415, flew with J35 engines instead of J46s due to Westinghouse's ongoing development problems. It was also built with F7U-1 landing gear due to the lead time required to develop forgings. Because of the difference in wheel sizes, both the main and nose gear doors had cutouts through which the tires protruded when the landing gear was retracted. (Vought Historical Center) F7U-1 BuNo 124422 was the backup carrier-suitability test airplane for 124415. (National Naval Aviation Museum) By agreement with BuAer, 415 still had the original speed brakes and did not have afterburners installed for the carrier trials. (NMNA)

Above, during shore-based catapult testing, NATC determined that the single-point attachment to the catapult shuttle was not as suitable for launch as a bridle that was attached to two points, one on either side of the fuselage. The new hook was located on the lower surface of the wing, just aft of the photo reference marking. Note also that the production airplanes have now been designated XF7U-1. (Author's collection) Below, F7U-1 BuNo 124415 was hoisted aboard Midway on 20 July 1951 for the Cutlass' at-sea carrier-suitability evaluation. (NMNA)

the throttle to idle. The airplane would promptly settle onto the deck in the middle of the arresting wires, one of which would be snagged by the tail-hook and pull the airplane to a quick stop.

Being too low or sinking too quickly would result in a ramp strike; being too high or not sinking quickly enough meant a trip into the barriers, sets of steel cables there to protect the people and planes on the forward half of the deck. Being slow risked a stall from which recovery before hitting the water was unlikely; too fast, floating up the deck and an abrupt and embarrassing stop in the barriers. Even the wave-off had its hazards: if power was applied too abruptly the result was either a roll and yaw that might not be controllable or an engine choking and losing power, with the plane stalling or descending into the water.

Naval aviators were well trained, however, and part of that process served to winnow out those who weren't up to the task. The short-final coaching of the paddle-waving LSOs was also a key factor in realizing a relatively low rate of crashes. Abrupt, wing-dropping stalls and poor handling qualities at low speeds were also rare in Navy carrier-based airplanes although not unknown. (The Vought F4U Corsair had its problems early on with the former and the Vought F8U Crusader notorious for the latter but both were tolerated because of their up and away performance).

There were differences between jets and propeller-driven airplanes that had to be taken into account. For one thing, without a propeller's immediate effect on low-speed thrust and drag, jets did not accelerate or decelerate very quickly. A jet pilot who got

Above, the hold back was also changed from a long cable attached to the airplane at the same point as the original catapult hook to a much shorter one attached to the keel on the aft fuselage. Note that the nose landing gear strut has been extended for launch. (NA 80-G-432150) Below, the launch begins when the holdback fractures at a predetermined load only reached when the catapult begins to stroke. With hydraulic boost on and the landing gear down, the ailavator trim tabs were automatically positioned full nose-up to minimize the nose-down pitch resulting from a hydraulic boost failure on takeoff or landing. (NA 80-G-432154)

low and slow behind the carrier on approach was at much greater risk of crashing into or even below the ramp than he would have been in most propeller-driven airplanes. Similarly, a late wave-off might not be possible although at least the jet pilot did not have to worry about torque, P factor, or pitch trim in the process, only whether the engine would spool up in time. Because of the lack of deceleration, when waving a jet, the LSO had to signal a cut when the airplane was farther out, since the jet did not sink as quickly as a propeller-driven airplane did when the power was pulled off. By 1951, these differences were well understood in the U.S. Navy, which had taken its first jet to sea five years earlier.

The Cutlass, however, would be the first swept-wing jet airplane that the U.S. Navy had landed aboard a carrier and even though further F7U-1 production had long since been cancelled, the naval aviation community was very eager to have the first-hand experience. It would also provide an early look at any peculiarities or deficiencies that might have to be accommodated or dealt with in the F7U-3 program. One known idiosyncrasy was the response to a nose-up input at approach airspeeds: the wing lost a little lift when the up ailavator was first applied since it initially introduced negative camber in the wing, causing a decrease in lift before the additional angle of attack increased it. The result was at best a slower response to aft stick. While a pilot might not object to this characteristic in a descending approach to an airfield, or even notice it, the precision of a level approach less than 100 feet above the water and only 20 feet above the flight deck was affected. Altitude control was cited as one of the F7U-1's approach problems.

As customary, the at-sea trials were preceded by more than 100 hours of shore-based flight test with BuNo 124415, involving 101 catapult launches and more than 125 arrested landings beginning in September 1950. These flights provided the requisite familiarity of pilot and LSO with the performance and handling quali-

Above, the high angle of attack required for maximum lift is apparent in this picture of the F7U-1 shortly after launch. (NA 80-G-432-149) Below, to keep the LSO in sight while in the groove, Feightner was obligated to fly in a left wing down forward slip. It is apparent in the movie that this picture was taken while the LSO is signaling high immediately before the cut but cuts him anyway. (Author) Bottom, because of its high drag at low speed, it turned out to be hard to be too high on approach. Most of the at-sea approaches were made well above the target wheel height of 20 feet.The white longitudinal lines on the deck are for reference as to distance from centerline at touchdown. (NA 80-G-432159)

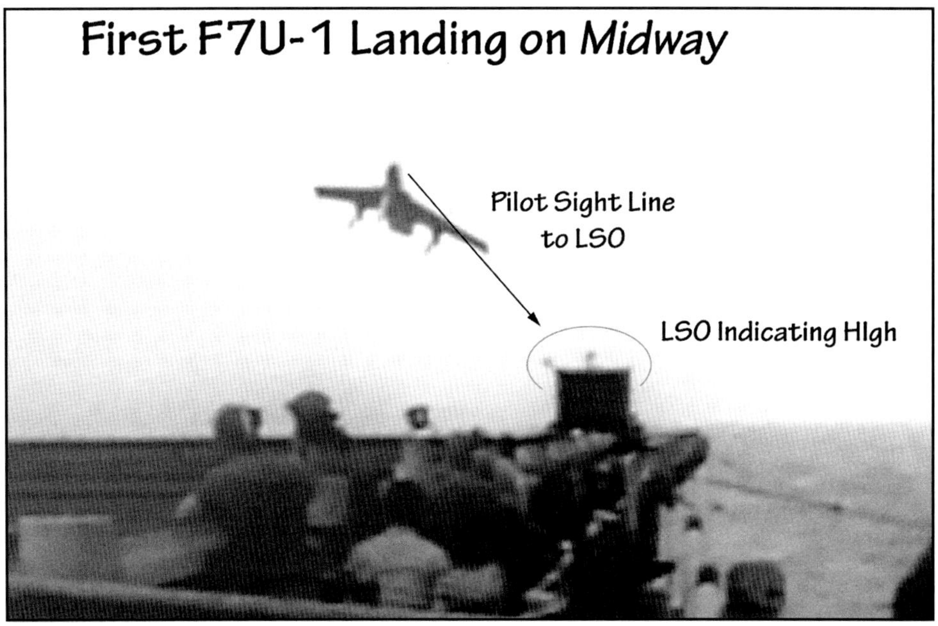

Above, the high approaches translated into high sink rates at touchdown, approaching and exceeding the design limit of the landing gear and airplane structure. (NA 80-G-432162) Below, on his second approach in the new Grumman F9F-5 Panther, Cdr George Duncan settled on approach and struck the ramp. (NMNA) Bottom, the F7U-1 was parked well forward when Duncan crashed but the program LSO was not out of harm's way. He was reportedly injured and the rest of the F7U-1's landings were waved by a substitute LSO. (NMNA)

ties of the aircraft prior to going aboard the ship. It had been a year-long process, much longer than usual, due to aircraft problems starting with the F7Us being grounded from 7 July to 24 August 1950 for the flight control system rework following the crash of 416. In addition, the aircraft was down for replacement of fuel cells and then a month to determine the cause of an in-flight vibration. The complicated fairing over the arresting gear failed three times during shore-based tests. It was eventually just removed along with the hook-shedding capability.

Once again, the limited view of the LSO was noted in the shore-based landings. A 26 April 1951 BIS "Yellow Sheet" recommended "That the contractor improve the pilot's view from the cockpit during the carrier approach. Arrested landing tests conducted with the model F7U-1 airplane have demonstrated that during the carrier landing approach, pilot view of the Signal Officer is extremely marginal when using the normal airplane approach attitudes of 17 to 22°." The identified problem was the distortion from the high degree of curvature of the Plexiglas side panel and the fact that the pilot was looking through it at a shallow angle. "Pilot vision during normal field landings is considered to be adequate. Pilot vision during taxiing is excellent."

Note that while the pilot's view of the LSO was termed "extremely marginal," he wasn't hidden except in crosswinds in excess of six knots from the left that generally wouldn't exist at sea. The carrier was generally headed into the wind for landings, perhaps with a slight crosswind from the left to keep the stack gas out of the groove.

In July 1951, Navy project pilot LCDR Edward L. "Whitey" Feightner was as ready as he was going to be and 415 was finally hoisted aboard *Midway* for at-sea trials. Since the Westinghouse schedule status hadn't improved, the engines were -34s, i.e. with no afterburners. Also aboard was a new Grumman F9F-5 Panther to be flown by CDR George C. Duncan for its at-sea qualification. Duncan made the first takeoff and landing, Whitey the second.

The launch was momentarily disconcerting when Whitey retracted the landing gear almost immediately after leaving the bow. The Cutlass started rolling to the left. He put the gear back down and regained control. Lesson learned, subsequent catapult launches proved to be satisfactory, even though 415 did not have afterburners. The first landing did not go well either, but unlike launches, there was no subsequent improvement.

On approach, the view of the LSO proved to be as bad if not worse than the original F4U Corsair, which was notorious and required a turning approach all the way to the cut. According to Whitey: "On the first landing, when I got in close, I didn't realize that I was skidding the airplane to keep the LSO in sight. He was almost abeam of me when he gave me the cut signal. When I took the cut and looked forward over the nose, I couldn't see any part of that carrier—nothing, not even the stacks—just water. I knew I had been cut late and I did what I had schooled myself never to do: I dipped the nose because I thought, I'm really going up the deck! I'll miss the barricade and go into the pack [of parked airplanes] up front! Well, instead, I snagged the number-one wire. The minute I dipped that nose, the airplane fell straight down. I barely made the flight deck! The tail hit eight feet from the ramp and scared the LSO to death."

Although Whitey described his approach as "skidding", what is apparent from the movie of that approach is that he is maintaining altitude in a forward slip, holding the left wing down with aileron and the nose to the right with rudder. He is also high, with the LSO signaling high immediately before giving him the cut, at which point Whitey straightens out and as he describes, lost all sight of the carrier until just before touchdown. He taxied forward and shut down for a discussion with the LSO.

Even if Whitey wasn't shaken by his first landing, what happened next undoubtedly gave him something to think about. On Duncan's second approach in the F9F-5, he started to settle in the groove and failed to add thrust in time. The result was a horrific ramp strike. Duncan was injured but recovered. His crash was attributed to a lack of recent experience with at-sea carrier landings: according to the accident report, he had not made one since 1944 but had "achieved adequate preparation for shipboard trials in (the) F9F-5 during carrier suitability trials at NAS PaxRiv."

Feightner's discomfort on his first landing followed by Duncan's crash probably affected his subsequent F7U-1 approaches. The carrier com-

Below, the F7U-1 was refueled before each takeoff to keep the landing weight constant. The first two landings were made at a weight of about 17,600 lbs; the next four were at a weight between 18,520 lbs and 18.920 lbs. The penultimate landing was at 17,820 lbs; the weight was not recorded for the last one. (NA 80-G-432151)

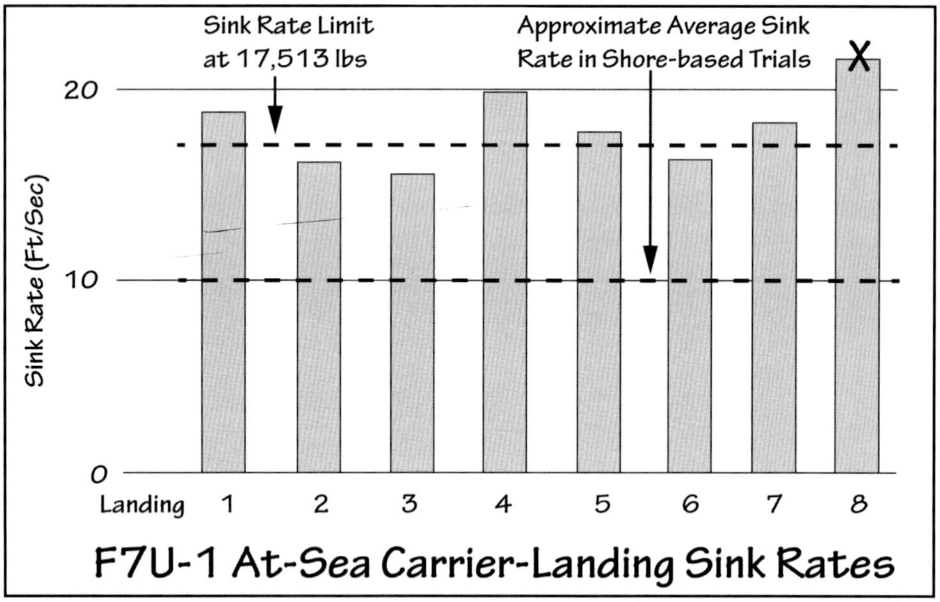

Above, most of the at-sea approaches were high. High sink rate landings resulted . However, the LSO was reluctant to signal a high correction once the F7U began to settle. At approach speed, it took more than aft stick to level out. (Author) Below, a crack was found in the forward fuselage after the eighth shipboard landing, terminating the at-sea evaluation. It was subsequently repaired and the airplane returned to flight status. (Author's collection)

---

patibility of the F7U-1 was still somewhat of a question mark. The F9F-5, on the other hand, was a derivative of the well-proven F9F-2. Ramp strikes were relatively rare, particularly in the benign sea conditions that existed for these trials. If the -5 had trouble coming aboard, then what was to be expected of the Cutlass? It also didn't help that the F7U Project LSO was injured by flying debris from Duncan's crash and a substitute had to be employed who was less experienced with the F7U's approach characteristics. Moreover, he, like Whitey, would now have been doubly apprehensive about too low an approach.

Whitey again: "We continued on with the trials, making landings 15 feet off centerline, left and right. On the last landing, the LSO was so nervous that when he gave me the cut, my landing gear was 33 feet above the flight deck! We had pictures to prove it. Not being able to see the deck, I just let it drop from there. I waited and waited to touch down, and just as I started to worry, blam, the airplane hit the deck, and it sounded as though somebody had taken a big flat board and clobbered the wings with it. The fuselage broke behind the cockpit and you could see daylight at the midpoint, but it didn't collapse on the deck and break off completely. That ended the project, but we still hadn't reached the yield on the nose wheel!"

The actual damage was more of a crack but enough to force an early conclusion to the at-sea evaluation. The F7U had to be ignominiously craned off the carrier back at Norfolk instead of being launched for a flight back to Patuxent River. At least it was still in one piece.

Of the eight landings Whitey made in the F7U-1 on *Midway*, five were at sink rates in excess of the design limit of 17 feet per second at 17,513 lbs and the other three were just under it. The average sink rate of the at-sea landings was significantly higher than for shore-based testing. Only two of the 132 shore-based landing sink rates exceeded the limit, indicative of the benefit of objects in the pilot's peripheral vision and the concern at sea about a ramp strike (don't be low) or barrier crash (get down quickly). Cut heights were double or triple that of the shore landing, The sink rate of that last landing on *Midway* was measured at 21.5 feet per second or not quite 15 mph. It was also the highest touchdown speed of the eight landings.

For some reason, the unacceptability of the field-of-view from an F7U-1 on a carrier approach was not fully appreciated by Feightner himself prior to the at-sea trials. Vought test pilots had made simulated carrier approaches and landings with the XF7U, guided by an experienced former Navy LSO, as early as 1949, but it doesn't appear that they noted a problem. It is possible that the Vought LSO was not properly positioned relative to the landing zone or the pilots unconsciously approached so as to keep him in sight. The Navy flight evaluation results were inconsistent: the deficiency was specifically identified during the June 1949 NPE and confirmed by a Vought visibility study; it wasn't raised as an issue as a result of the March/April 1950 flight evaluations. However, Vought engineers

had developed improvements, at least some of which were to be incorporated in the F7U-2, and taken the requirement into account with the original design of the F7U-3 forward fuselage; it was deemed to be acceptable at its mockup review in April 1950.

My guess is that the shortcoming originally stemmed from a BuAer specification for over-the-nose visibility that did not take into account the very high fuselage-pitch at approach speeds of tailless airplanes like the F7U or even the F4D, which was modified as a result of its mockup review. The degree of the problem wasn't fully appreciated in the field landings accomplished because the pilot could still see enough of the 150-foot wide runway and the airport environs to center the airplane over the runway and know where its threshold was. (Another reason might have

Above, BuNo 124415 easily fitted on the deck edge elevator on the USS Midway on 25 July 1951. (USN) Below, LCDR Freightner taxiing in 124415 aboard CVB-41. Slats were lowered and area under slats was painted red. Tip of nose was off-white with white stripe on the left side only. (National Archives)

Above, in order to improve the pilot's vision of the deck and the LSO, the F7U-3's radome was reshaped and the windscreen was modified. (Vought Heritage Center) Below, Vought engineers updated the F7U-3 polar vision evaluation after the F7U-1's at-sea trials. It was now apparent that the view of the deck on approach was only acceptable bu F4U standards. (Author)

been that the pilots were accustomed to having relatively poor forward visibility in propeller-driven fighters. "Pete" Aurand, the commander of the Navy's first West Coast carrier jet squadron, comparing the excellent visibility of the North American FJ-1 Fury to the F4U Corsair's, noted that he could only see the LSO and a small part of the deck thorough the left quarter panel of the windscreen of a Corsair).

The major impact was on the F7U-3 program. Whitey flew to Texas a week or two after 425's at-sea trials to provide Vought's management and engineering with a first-hand report of the unacceptability of the F7U-1's over-the-nose visibility and to reevaluate the F7U-3's.

Vought redid the vision study as a result of the *Midway* trials, comparing the original F4U, F7U-1, and F7U-3 views of the deck. As before and confirming Feightner's unhappy experience, it showed that no part of the carrier could be seen from the F7U-1's cockpit by the pilot in the normal seated position when on short final, not even the ship's mast. The F7U-3 was only marginally better. BuAer requested a redesign to further increase the pilot's view of the carrier on approach.

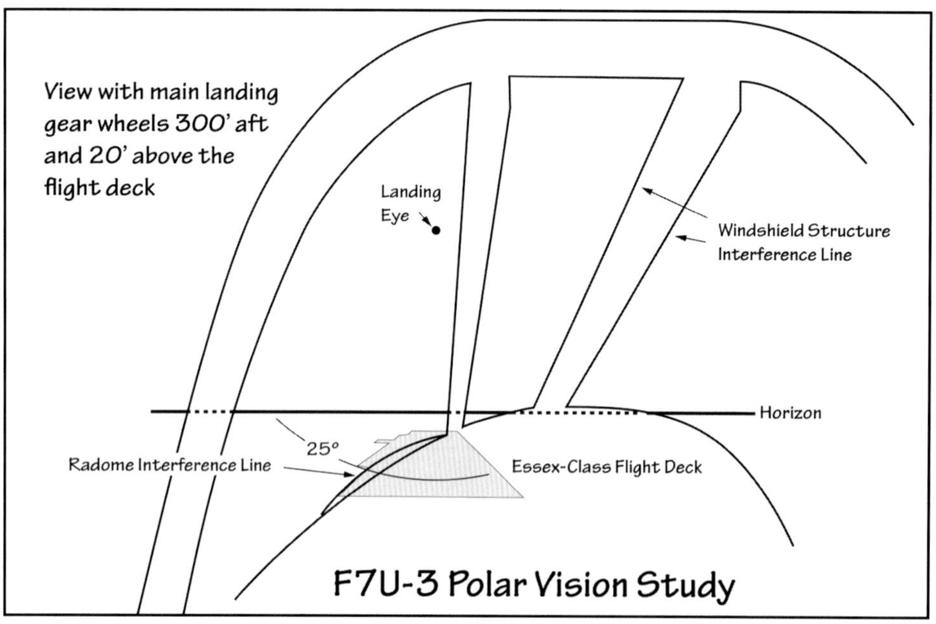

As a result, the F7U-3 forward fuselage was reshaped from the seat bulkhead forward. The bulbous radome was replaced with one that had a downward slanted upper surface. The windshield and sliding canopy were modified for a higher seating position. The ejection seat was modified to provide a raised and tilted position for landing. Mockup reviews of the new configuration were accomplished in early October and mid-December 1951.

It was too late to incorporate the changes in the first F7U-3. It was in final assembly and its first flight was only a few months away. By herculean effort they were incorporated in the fourth F7U-3 in time for its scheduled carrier suitability testing. The production effectivity was F7U-3 number 17.

If VX-3 operation hadn't already been cancelled, it surely would have been following Feightner's evaluation of 415 at sea aboard *Midway*. Further at-sea operation of the F7U-1 was now considered to be "extra hazardous in the present configuration." However, BuAer requested that Vought investigate the possibility of increasing the visibility of four airplanes at minimum cost and calendar time. Vought informally suggested that the upward seat travel could be increased a few inches and an airstream deflector added to the top of the windscreen (at the time, it was a standard practice to take off and land from a carrier with the canopy open to facilitate egress in the event that the airplane had to be ditched). After due consideration, the Fighter class desk decided that little benefit would be derived from VX-3's operation of the F7U-1s.

Stability and Control Tests

BuNo 124419 was received by Flight Test at Patuxent River on 10

April 1951 for the Stability and Control test portion of the Service acceptance Trails. Progress suffered from excessive maintenance due to aircraft and instrumentation failures.

Following instrumentation and maintenance, flight test began on 6 July 1951 and continued through 10 August, when the aircraft was downed again for maintenance. Flight test resumed on 12 September and continued through 31 October, when the airplane again was downed for two weeks for maintenance.

Royal Navy test pilot Eric "Winkle" Brown was assigned to NATC as the Resident British Test Pilot in 1951/2. He was already well on his way to having flown more different aircraft types than any other pilot, in most cases writing about their performance and handling qualities candidly and in detail. His record of 2,407 carrier landings is unlikely to ever be broken.

On 21 October 1951, Brown put BuNo 124419 through its paces. He was not impressed: **"It had a tendency (on the ground) to weathercock in even moderate side winds. Rudder control effectiveness in a crosswind take-off was unsatisfactory until a speed of 60 knots was reached. Up to that point, directional control was dependent on the use of brakes.**

The climb was laborious and there was a distinct immediate impression that the aircraft was underpowered.

The longitudinal stability was unsatisfactory at high indicated airspeeds and at high altitude and the longitudinal control was too sensitive near the neutral position...The rate of longitudinal feel trim at slow speeds was also not fast enough for comfort. Lateral control force stability was noticeably synthetic and the "springyness" associated with stick centering was objectionable. The strong stick centering was poorly damped and this, coupled with the sensitivity of the lateral control near the neutral position, caused overcontrolling and made precision flying difficult. Lateral trim characteristics were unsatisfactory. It was difficult to make small adjustments in lateral trim because of the sensitivity of the lateral control system.

There was insufficient ailavator to stall the aircraft except dynamically, when a wing drop usually resulted.

The catapulting trials at Pax River reveal the large amount of longitudinal rotation required to develop maximum lift at the end of the catapult power stroke.

Deck landing...presented a problem because of the very poor view caused by the approach at a speed of 110 knots, which put the aircraft at 20 degrees incidence."

Brown did like some Cutlass features: **"The brakes were power boosted, giving low pedal force and positive action...It was quite easy to taxi because of the powerful brakes.**

The rate of roll was excellent, being 250 degrees per second at 250 knots.

The yaw damper was most effective in controlling the aircraft's Dutch rolling tendencies in turbulent air.

The speed brakes were very effective in reducing the speed of the aircraft and they produced no buffet.

Handling characteristics with asymmetric engine power were very satisfactory in normal flight and were reasonable for airfield landing."

His summary was scathing:

> **"The XF7U-1 Cutlass was, in effect, a total failure. It was underpowered, overcomplicated, and a non-starter for carrier operations. I regard it as a test bed for the irreversible powered control system with completely artificial feel and in this case, not a very successful application. The system found favor with designers because it permitted maneuvers at high sonic, transonic, and supersonic speeds and also it masked stick fixed instability, particularly in the transonic range. This second reason is a dan-**

Below, F7U-1 BuNo 124419 was used to accomplish the Stability and Control BIS evaluation. Vought's decoration on the nose was removed and FT for Flight Test and 419 were painted in its place. For some reason, the white checkerboard pattern was also added to the inboard side of the fins about the time that it was transferred to NATC. (NMNA)

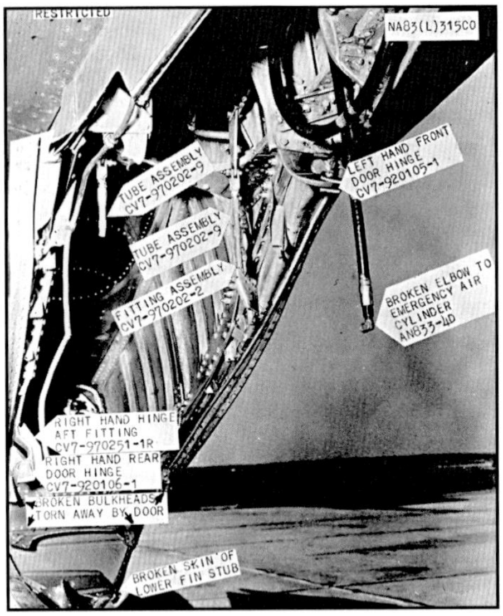

gerous one, since failure of either the power system or the artificial feel system may lead to the abrupt application of large control surface deflections at high speeds, with attendant high aerodynamic forces and moments—a recipe for disaster."

A month later, on 19 November, a severe pilot-induced longitudinal oscillation did in fact occur during a Navy test flight of 419 to evaluate trim changes with speed brake extension. The first test point was at 430 knots at 10,000 feet, which was reached in a shallow dive. The pitch and yaw dampers were off. When the speed brakes were opened, the nose pitched down. An unexpectedly large pitch up of more than 6 G (the recording needle went off the chart) followed when the pilot countered with aft stick. Recovery from that produced more than -6 G, again off the chart. The longitudinal oscillations continued as the pilot closed the speed brakes and the throttles. They took six more cycles to damp out, with peaks of plus 6 G or more and minus 2 to 3 G. That the only structural damage was the left main landing gear doors were torn away was a testament to the strength of the Cutlass airframe.

The repairs were easy. The troubleshooting of the cause was not. After two months of investigation at Pax, including replacement of some of the components in the longitudinal feel system, the cause remained unknown. The aircraft was therefore ferried back to Vought in mid-January 1952, restricted to a maximum indicated airspeed of 475 knots or Mach 0.92. The planned high Mach number tests were postponed until Vought could determine the cause of the incident and eliminate it.

As it turned out, the mystery was solved on the ferry flight back to Texas. The pilot noted that the stick occasionally snapped forward during cruise, leading to the discovery of a malfunctioning proportional pressure valve, which was supposed to increase the hydraulic pressure to the feel system as speed increased. Vought engineers therefore deduced that the incident was caused by a malfunctioning longitudinal-feel system combined with a proportional pressure valve that occasionally stuck. The dual failure, coupled with the pilot moving the stick faster than the ailavator power cylinders could respond when the failure occurred, resulted in the violent pilot-induced oscillation in pitch.

The incident caused yet another disruption in Vought's flight test planning, as 419 was supposed to be complete its stability and control BIS trials on or about 22 November. It was then to be returned to Vought for modification and about five weeks of Vought flight test before being returned to the Navy. It finally got back to Patuxent River in early March for completion of the required structural demonstrations.

In May, it suffered another incident during a steady roll during the completion of the Contractor's Structural Demonstration. On the first or second of three continuous rolls, the left wheel well doors were torn loose again. The aircraft was again ferried back to Vought for installation of redesigned doors.

Although the Navy test pilots in their final BIS report on Stability and Control mirrored all of Brown's findings with respect to handling qualities, they were not nearly as critical, generally rating them as acceptable to satisfactory. In some cases, they considered something that he couched as a shortcoming to be a feature. For example, in noting that the ailavators did not have enough control power to stall the airplane in

Above left, although the F7U-1 pilots occasionally exceeded structural limits without untoward consequences, the landing gear door locks proved to be a weak link. In this instance, it was during a structural demonstration of three consecutive rolls. (USN, National Archives) At left, BuNo 124419 with FT for Flight Test on its nose at an open house at NAS Patuxent River, MD. (Ginter collection)

level flight and stalls were benign in accelerated flight, the report recommended removal of the Safe Flight stall-warning device. In any case, it was apparent that the Navy did not consider any of the unsatisfactory conditions to be irremediable.

Taking Flight With the Angels

In 1951, the Admirals were feeling some heat about the Navy's lack of operational swept-wing fighters while MiGs and Sabres were dueling in the skies over Korea. By mid-year, the only one of its high-performance jets that had flown was the F4D Skyray, and that with an interim engine. The XF3H and XF10F were hamstrung by delays with the Westinghouse J40 engine. North American and Grumman had contracts for the FJ-2 and F9F-6 respectively but, like the F7U-3, they would not fly until later in the year.

In late 1951, probably desperate to showcase something, the Admirals agreed to assign two F7U-1s to the newly reinstated Blue Angels, augmenting their air show performance with the latest version of the straight-winged F9F Panther. BuNos 124426 and 124427 were painted the unique light blue of the team and given the tail numbers 7 and 8 respectively. LTCDR Edward L. "Whitey" Feightner, who had been the F7U-1 Project Officer at NATC Patuxent River and was therefore the Navy pilot with the most flight time in the Cutlass, was to fly one and LT Harding C. "Mack" McKnight, a former Blue Angel, was chosen to be the pilot of the other.

The two airplanes assigned to the Blue Angels had been modified to have the all-metal vertical fins strengthened with the external doubler required following the failure of a fin in flight on 420. To demonstrate that they were strong enough, Vought test pilots made full-stick rolls at 550 kts and 3 gs; 6 g pullouts at 500 kts followed by an Immelmann turn; full-stick continuous rolls at 500 kts; and a 4.5 g full-stick rolling pullout at 300 kts.

Due to the ongoing trouble with Westinghouse engine deliveries, non-afterburning J34-WE-34s were installed, the same engines as in the F2H-2 Banshee. Since the F7U weighed at least 2,000 lbs more than the F2H, the lack of afterburning resulted in a lower thrust-to-weight ratio than the fairly pedestrian Banshee's. To partially compensate, Vought stripped these two aircraft of all unnecessary hardware like wing folding actuation, the catapult hook, armament, nose strut extension, etc., in order to reduce their weight. Nevertheless, takeoff and climb were still not nearly as impressive as they would have been using afterburners. The exotic swept and tailless configuration and high roll rate had to suffice as crowd pleasers.

As it happened, the new F9F-5s that the Blues were to fly were grounded at the start of the 1952 season by a fuel control problem. The arrival of two F7U-1s in January at NAS Corpus Christi, then the home of the Blue Angels, was therefore more welcome than it might have been otherwise. For the first scheduled shows of the season, the Cutlasses were all the Blues could put up.

It's not clear how many shows the F7Us actually flew in. According to an interview with Feightner, he flew the Cutlass in six shows between March and June 1952. Although Feightner and McKnight almost certainly practiced their demonstration at Corpus Christi, there's no record of a formal show there. Feightner does describe a show put on for the Texas Association of Student Councils at NALF Cabaniss that almost ended in disaster due to the sudden appearance of a light plane in literally the middle of his routine. In published interviews, Feightner also describes a near-crash at Pensacola and mentions an appearance at Pittsburgh.

Below, in late 1951, Vought prepared two F7U-1s, BuNo 124426 and 7, for assignment to the Blue Angels. All unnecessary equipment was removed and the aircraft were painted and marked exactly like the Blues' existing aircraft. (Vought Heritage Center)

Above, after the F7Us joined the team, the Chance Vought Cutlass logo on the outboard side of the fin was replaced with a 7 (426) or 8 (427). BuNo 124426 as seen in June 1952 at NAS Glenview, IL. (Clay Jansson) Below, because of the ongoing problems with the qualification of afterburning J34s, the Blue Angels F7Us were powered by non-afterburning engines. The fins were reinforced with an external strap that spanned the attachment of the upper fin to the lower. BuNo 124427 is seen here prior to receiving the number 8 on its twin fins. (NMNA) Bottom, the markings of the F7U-1s assigned to the Blue Angels included the large U.S. Navy on the underside of both wings. This picture was taken when the Navy's Gound Controlled Approach Unit 16 at NAS Corpus Christi was guiding a Blue Angels F7U for the unit's 40,000th GCA approach. (Vought Historical Center)

They arrived, somewhat colorfully, for a scheduled show at NAS Glenview, Illinois, in early June and left the F7Us at their last stop, NAS Memphis, Tennessee, later that same month but it appears that the Cutlasses were only on static display at both.

According to Feightner, the first show the Cutlass was in was at Pensacola's Saufley Field in front of the Secretary of the Navy and 500 Navy supporters in the business community. Since the weather was gusty and McKnight only had a total of five flight hours in the F7U, Whitey elected to make it a solo demonstration. He made a maximum performance takeoff and while he was climbing out, the hydraulic system quit. Literally out of control while the automatic switchover was taking place, he rode along while the airplane peaked in the climb, pitched over, and started down. The controls became effective in the manual mode just in time for him to be able to pull out and miss hitting the ground but not the trees at the end of the runway. They relieved him of a portion of the leading edge slats, a piece of which entered the left engine and caused it to fail, requiring a single-engine landing. At the Secretary's request, he then got into the second F7U-1 and flew the planned demonstration.

It was not an auspicious start for the F7U's short air show career with the Blues but it proved all too typical. Four J34s had been allocated to the unit for spares and at least that many were used. In-flight emergencies and excessive maintenance for component and system failures were the rule. The absence of hydraulic fluid leaking out meant that the system was fully depleted. At the conclusion of one flight, when McKnight put his wheels down he discovered that one main landing gear was positioned with its wheel in the takeoff (forward) position and the other, properly angled aft. However, he had no problem landing it that way.

The F7Us appeared in conjunction with the opening of the Pittsburgh International Airport on 31 May 1952, although their presence was limited to a flyby on arrival and static display. On departure for the next venue, NAS Glenview, Whitey had an engine failure but elected to continue on because of the poor weather at Pittsburgh. As they came overhead at Glenview, McKnight's right engine exploded and his left engine flamed out. He made a successful emergency landing but his airplane was now blocking the runway and Feightner was low on fuel. He was directed to land at nearby O'Hare Airport. It was still under construction and not yet open for aircraft operations so he was the first to land there.

Both airplanes were again repaired and returned to flight status, but not for long. While they were still at Glenview, the Chief of Naval

Aviation Training sent a telex to the CNO that noted the number of engine, fuel control, and hydraulic failures was excessive and there were insufficient spares to support the airplanes. "Possible good will to be generated by having F7U as part of Blue Angel team more than offset by probable failure to meet schedules." Moreover, now that the Blue Angels were finally ready to resume flight demonstrations in their Panthers, the number of enlisted personnel assigned was inadequate to support both aircraft. It concluded "Continued flying of F7U-1s by this command considered to serve no useful purpose."

The first scheduled Blue Angel appearance with the Panthers was in June at NAS Memphis, then the home of the Naval Air Technical Training Command. The two Blue Angel F7U-1s were flown there in time for the show but as had happened all too frequently, arrived with maintenance issues. They apparently did not participate in the flight demonstration and probably never flew again. Instead, they remained at NAS Memphis as ground-bound maintenance trainers.

March 1952 Air Force Evaluation

Vought offered derivatives of the F7U to the Air Force at least three times. In September 1949, as an alternative to the Navy smoothing Vought's production manpower with the addition of 10 more F7U-1s, Vought proposed them to the Air Force, with deliveries to begin in January 1951 at three per month, the 10th airplane to be delivered in April. The Air Force configuration would not include "the easily removable portions of the arresting gear, catapult gear, wing folding and barrier guard." The price was $500,000 per airplane, not including Government Furnished Equipment (e.g, engines and radios), spare parts, special tools, and ground handling equipment. Vought's proposal also pointed out that if the Air Force was interested in a greater quantity of aircraft, the government-owned, Air Force-managed plant at Kansas City would be quickly outfitted to produce as many F7Us as needed. The Air Force passed.

Also in 1949, Vought promoted its V-366 derivative of the F7U-1 as an interceptor to the Air Force. It was to be powered by two J46-WE-2 engines and was only about 1,000 lbs heavier than the F7U-1. Once again, the Air Force declined to add to its already full plate of fighter development programs.

In early 1952, Vought proposed its V-377, a modified F7U-3, as a fighter-bomber to the Air Force, which this time requested a flight evaluation by its pilots of the F7U-1. Vought was only given a few days notice. As it happened, the four available F7U-1s at NAS Dallas had different engine and flight control system configurations, given the ongoing Westinghouse engine problems and the flight control system refinement in progress. As Paul Thayer, now Vought's Vice President of Sales, told the visitors, "We have some (Cutlasses) that dive well, some that climb well, and some that have good handling characteristics."

Aircraft #4, BuNo 124418, had one -42 engine with an afterburner and one -34 without; it had been flown as high as 40,000 feet. However, it was structurally limited to 4 G below 400 knots and only 3 G above. It had the best feel system of all the F7U-1s, incorporating centering springs on the valves, but it did not have a pitch damper installed and the yaw damper was malfunctioning.

Below, BuNo 124418 participated in the Air Force evaluations of the Cutlass. It is seen here on static display at Dayton, OH, on 5 September 1953. U.S. Navy on the forward fuselage was white.(Bill Steeneck via W.T. Larkins)

Aircraft #5, BuNo 124419, had -34 engines without afterburners. As a result, it was 1,000 lbs lighter than #4 but had relatively poor climb performance, with climbs to 30,000 ft and above "a struggle." On the other hand, it was structurally limited to 6 G up to 500 knots and 4.5 at higher speeds and it had operative pitch and yaw dampers.

Aircraft #6, BuNo 124420, was configured the same as #5 but was considered to have a better feel system.

Aircraft #8, BuNo 124422, had -34 engines with Solar A-100D afterburners installed. It had the best low altitude performance of the four but hadn't yet been flown in afterburner above 20,000 feet. Its feel system had also not yet been fully checked out and adjusted but it did have operative pitch and yaw dampers.

The very experienced Air Force test pilots were GEN Albert Boyd, commander of the newly-created USAF Flight Test Center at Edwards AFB; COL Alfredo "Fred" Ascani, his vice commander; LtCol Richard "Dick" Johnson, who had set a speed record in an F-86A in September 1948; and LtCol Frank "Pete" Everest, who was to take the Bell X-2 out to Mach 2.9 in 1956. The F7U-1 evaluation consisted of 15 flights and a total of 8.5 hours. North American F-86 Sabres were used for chase and pace.

GEN Boyd made the first flight of the evaluation, flying 418. Unfortunately, the cloud ceiling was only 2-3,000 feet so he couldn't explore much of the flight envelope. However, he made several afterburner starts and stated that the F7U-1 was a "very easy airplane to get used to."

Even in afterburner, the USAF pilots noted "this aircraft suffers considerably from inadequate thrust in all flight conditions." In 422, which had the Solar afterburners, COL Ascani accelerated to Mach .92 in level flight at 20,000 feet in afterburner but performance at higher altitudes was disappointing. He stated that there was not much thrust apparent from the afterburners above 30,000 feet.

COL Ascani subsequently took 418 up to 40,000 feet and dove it to Mach 1.02, "booming" North Dallas, which resulted in several rattled windows and citizens. Full thrust and a 60-degree dive angle were used. He used the speed brakes in the recovery from the dive and stated that they were very satisfactory—as well as essential since the ailavators could only produce a 2 g pullout at supersonic and transonic speeds.

Not to be outdone, LtCol Johnson also boomed NAS Dallas in 418 along with LtCol Everest flying the F-86. Johnson stated that the F7U went supersonic "better than the F-86," at least from a handling qualities standpoint. A pitch down occurred at Mach .91 but the trim change was gone by Mach .95.

There were other qualitative comparisons with the F-86. The F7U-1 could fly slower but not turn as well, pull less g at altitude before experiencing heavy buffet, and lost more speed in hard turns. The large, relatively thick wing was a handicap. For the same reason, it was as fast as the Sabre in a dive up to about Mach .83, but then the lower critical Mach number and corresponding drag rise of the thick Cutlass wing began to tell.

The Air Force pilots were complimentary about some characteristics:

I. Taxiing and ground handling were superior to those of existing USAF jet fighters.

Below, BuNo 124419, shown here flying by downtown Dallas, Texas, was one of four F7U-1s flown by the Air Force during its March 1952 flight evaluation. One of the Air Force pilots used another F7U-1, BuNo 124418, to "boom" Dallas. In those days, sonic booms were considered "the sound of freedom." (Vought Historical Center)

II. A noteworthy feature to be considered in future planning was the power-boosted brake system. Brake action was positive and only low pedal forces were required.

III. Visibility while taxiing was excellent and was superior to that of any USAF jet fighter now in use.

IV. Rate-of-roll was considered satisfactory, possibly in excess of that required.

V. When performing accelerated maneuvers, no noticeable tendency to over-control nor were any unusual or undesirable characteristics noted.

VI. In comparison to other aircraft flown in high-speed dives, this aircraft exhibited excellent handling qualities in the high Mach region, with the exception of the pitch-down deficiency encountered at 0.9.

The rate-of-roll comment would appear to be an understatement. NACA measured the maximum roll rate of an early F-86 to be 138 degrees per second. In Navy tests, the maximum F7U-1 roll rate was measured at 250 degrees per second.

Their conclusions were wholly negative in tone with respect to the fighter-bomber requirement, specifically:

I. The longitudinal trim change evidenced as a tendency to pitch nose down, which is encountered at 0.91 indicated Mach number, is unacceptable for a fighter-bomber type airplane.

II. The lateral-directional dynamic stability is not satisfactory for a gun platform, even with the presently installed damping device operating.

III. Longitudinal control is restricted or limited by the light stick force per "g" gradient and by the decrease in elevon power or increase in stability encountered at high Mach numbers.

All, of course, were being addressed in the F7U-3 program. It seems likely that the Air Force would have passed on the V-377 in any event as not being superior to any of the fighter-bombers that they had in development.

Above, BuNo 124420 at Dayton National Air Show in 1953 where LTCDR Puckett put on a spectacular air show. (Robert L. Stuckey via W.T. Larkins) Below, BuNo 124422 was another F7U-1 used in the Air Force evaluation. Note the two long chutes mounted on the bottom of the forward fuselage that channeled the cartridges and links past the engine inlets before they were dumped into the free stream. (Vought Historical Center)

Spin Test Program

Vought test pilots evaluated the spin characteristics of the F7U-1 in a two-week program between 23 July and 8 August 1952 at Vought. BuNo 124420 was used for the spin testing. For the spin test program, modified outboard wing panels were substituted that had a canister on each wing tip loaded with a 64-inch diameter spin-recovery parachute. The lower surface of each wing was striped with yellow paint to aid the chase pilot in counting the number of turns in each spin and determining the F7U attitude. Before the spin evaluation was begun, the spin–recovery parachutes

Below, for the spin test program, F7U-1 BuNo 124420 was configured with spare XF7U-1 outboard wings modified with spin-recovery parachute canisters on the tips. (NMNA)

Above, for the spin-test program, a spin-parachute control panel was mounted in place of the gun sight. (NMNA) Below, the spin-recovery parachutes were small but they were acting on a long lever arm. They were tested in flight but never needed during the spin demonstration program. Unfortunately, the F7U-1 proved to be not representative of the bigger, heavier F7U-3 from a spin recovery standpoint. (NMNA)

were deployed at 157 knots EAS and jettisoned to insure proper function if needed.

A total of 16 power-on and power-off spins were accomplished. The attitude of the airplane in the spin was 50° to 70° nose down. Rotation was slow and approximately every one-half turn the airplane exhibited a tendency to recover even though full pro-spin controls were being held. No difficulty was encountered in recovering quickly from any spin. Recovery required nothing more than relaxing the pro-spin control input to neutralize the control surfaces. Control forces required for recovery were negligible.

The aircraft was then ferried to Patuxent River for the formal spin demonstration, which was accomplished in late August and early September 1952. The take-off gross weight was 21,560 lbs and the cg was at 17.9% of mean aerodynamic chord. The maneuvers and results were:

1. Roll-out from inverted flight at 126.9 knots by full lateral stick motion: no stall/spin resulted
2. Power-on spin to the right entered from a 2.3 "g" accelerated stall and continued for 5 turns: recovered in 1/8 turn
3. Power-off spin to the left with stick 10 degrees against the spin: could maintain only 2 turns before the aircraft recovered on its own.
4. Power-off spin to the left with stick 1/3 with the spin and continued for 5 1/4 turns: recovered in ¼ turn; 11,500 feet of altitude was lost at an average rate of descent of 300 feet per second.

As in the tests at Vought, spins that did develop and could be maintained were recovered from by just neutralizing the controls. There was some damage to the rudder control mechanism and spring tab due to the heavy buffeting that was associated with the spins as the number of turns increased, but no corrective action was required since a spin of more than one or two turns was considered unlikely.

As it turned out, this evaluation did not provide an adequate predic-

tion of the F7U-3's post-stall characteristics.

I Feel the Need, the Need for Speed

By mid-1952, Vought was completing its development of the F7U-1. The traditional uses for preproduction airplanes were winding down. However, it was the fastest airplane that the Navy had, not counting the D-558 research aircraft at Edwards and the F4D and F3H prototypes. After the Blue Angels' Cutlasses were retired at Memphis, there were still 10 available for assignments that took advantage of that.

At that time, all but 124428 were still involved in program efforts: three were assigned to the truncated BIS evaluation (124415 and 124419) and spin demonstration (124420); three were involved in F7U-3 program support, two for Westinghouse engine development at Vought (124424 and 124425), and one for catapult tests (124421); and three were finishing up miscellaneous Vought tests (124418, 124422 and 124423). In July, BuAer anticipated that 124418, 124421, 124422, and 124423 would be assigned to Tactical Test at NATC in the fall; 124419 and 124420 would be operated by Vought for "Regulus Control" and 124428 would be assigned to NACA Langley in August for "Research." BuNo 124415 would be finishing up the BIS evaluation and 124424 and 124425 would continue on with Vought support of Westinghouse engine development.

The October assignment plan changed slightly. 124422 was now to be assigned to the Naval Air Missile Test Center (NAMTC) at Pt. Mugu on 1 December for Sparrow missile development instead of Tactical Test at NATC. It would be replaced by 124415 when BIS evaluation requirements were fulfilled. The NACA assignment date had slipped to November.

By November, the NATC Tactical Test effort (tactical value of augmentation of engine thrust by use of afterburner, interceptor capabilities of the F7U design, and direct comparison of with available contemporary fighters) was reduced to two airplanes. One of

Above, Vought flight line on 3 January 1952. From left to right, the first F7U-3, BuNo 128451, and F7U-1s BuNos 124426, 124424 or 5, 124422, 124427, 124418, and 124420. The F7U-3 had just flown two weeks before; 426 and 427 are being prepared for the flight to NAS Corpus Christi to join the Blue Angels. (Vought Heritage Center) Below, two F7U-1s were assigned to the Navy's Missile Test Center for support of Sparrow missile development. Note that vertical fin on BuNo 124422 has been modified with the external reinforcement strap. All the cockpit access steps have also been opened. It appears to be ready for or just returned from a flight demonstration at an air show at which the Blue Angels appeared with their F9F-5s and the newly formed U.S. Air Force Thunderbirds, with F-84Gs. (Jay Miller Collection)

the reductions was necessary to provide an F7U-1 to Convair (Consolidated Vultee) to provide high-speed chase for its new XF2Y-1 Seadart.

In December, the plan was for the 10 F7U-1s still on flight status were to be assigned as follows:

Location	Qty.
Vought **for Westinghouse engine development**	**2**
NATC **Tactical Test (TT)**	**3**
NAS Pt Mugu **to Support Sparrow and Regulus development programs**	**3**
NACA Langley **Research**	**1**
BAR San Diego **Bailment to Convair for XF2Y-1 chase**	**1**

One of the prospective Navy mission assignments for the F7U-3 was fleet air defense against enemy bombers. The armament was to be up to four radar-guided Sparrow missiles. The F3D was being used for the Sparrow development program. The F7U-1 was to provide an early capability to measure the aerodynamic loads on the Sparrow I missile and launcher at high-speed, something the F3D would have been limited at since it was powered by the same engines (without afterburning) but was much larger and heavier. Pt Mugu had developed a "captive balance device" for this purpose. It was essentially a section of the Sparrow equipped with special internal structure and strain gages that measured the pitching, yawing, and rolling moments acting on the missile when it was carried in flight and in positions simulating the first two feet of missile travel after launching. BuNo 124422 was ferried to Pt. Mugu in late October 1952. The installation of the launcher and instrumentation was in progress by the end of the year and nearing completion in mid-January 1953.

There were also plans to modify and reinforce a spare F7U-1 left-hand outer panel to add a pylon. In January 1953, Vought quoted design and fabrication of "a high fineness-ratio pylon using Sperry-furnished Douglas jettisoning mechanism and adjustable forward and aft support fittings" for the left-hand wing center section as well as the existing inner pylon location. It seems likely that the benefit was not considered worth the cost, since the Sparrow-armed F7U-3M was to fly in 1954.

As planned, 124421 was ferried to San Diego in January 1953 for bail-

Below, although the F7U-1 apparently wasn't used to chase the Convair F2Y-1 as originally planned, it was used upon occasion to chase the F7U-3. LCdr John W. Ryan, OINC Flight Test BuAer, flying BuNo 124418, accompanied an F7U-3 to NAS Kingsville, Texas, in April 1953 for its armament testing. (NA 80-G-480928)

ment to Convair. The XF2Y Seadart first flew in April but was chased on its first flight with a Douglas AD-5. John Knebel, a Convair test pilot who was to chase the Seadart in it, remembers it as "a massive hydraulic leak" and hated by maintenance. He never did fly it.

In any event, BuNo 124421 was ferried from San Diego to Pt Mugu in August 1953 to join 422. It was removed from flight status in September 1953 and formally stricken there in January 1954, apparently having flown little, if any, according to its Aircraft History Card. It was probably used as a source of spare parts for 422, which was flown about 40 hours over the two years that it was based at Pt. Mugu.

It is likely that the Cutlasses on the west coast were shortchanged from a spares standpoint due to the redesignation of the production F7U-1s as XF7Us back in June 1951. Initially, unbeknownst to BuAer, the Aviation Supply Office (ASO) immediately took action to cancel procurement of spare parts for the F7U-1 program, since strictly speaking, spares support of experimental aircraft was the responsibility of BuAer. BuAer clearly assumed that ASO would continue to procure spares since the funds for that purpose allocated to ASO had not been taken back. BuAer had, for some reason, redesignated the XF7U-1s as F7U-1s in November 1951. However, ASO didn't advise BuAer of its action until requests were received from the Blue Angels for spare parts in January 1952.

The ASO seemed to be on firmer ground than BuAer, bureaucratically: "It was assumed that BuAer would provide future support for the XF7U-1 experimental aircraft in accordance with BuAer Manual of Program responsibilities." Moreover, in its January 1952 memo to BuAer, notwithstanding the redesignation and the fact that it still had the funds for spare parts, it insisted "Responsibility for support should, therefore, continue to be vested in the Bureau of Aeronautics." In any event, the damage to the F7U-1 program was done. Spare parts generally had a lead-time of 12 months.

In November 1954, 422 was either ferried or trucked to the Overhaul and Repair facility at San Diego and "retired"; in June 1955 it was formally stricken and transferred to Lakehurst for whatever test benefit it could provide there.

The use of F7U-1s for Regulus missile control was dropped in favor of using the existing control airplanes and then the Grumman F9F-6 Cougar in that role. The Cougar wasn't as fast as the Regulus but it was operational in the fleet. The cost of modifying two F7U-1s and the spares situation were doubtless factors. It's possible that the December plan anticipated basing one or two F7U-1s for Regulus safety chase since they were fast enough to stay with a rogue one and shoot it down.

The last production F7U-1, BuNo 124428, was ferried from Vought to NACA Langley in December 1952. According to its aircraft history card, it was assigned to Philadelphia in mid-March 1954, more than a year later, and stricken in June. No flight test reports were found in a search of the Langley archives. In August 1953, Langley requested disposition of the spare -42 engines it had received along with the airplane. That and the recorded total time of 22 hours suggests that it wasn't flown much, if at all, while there. However, any NACA flight time might well have not been included in the Navy's total.

The Tactical Test (TT) F7U-1 evaluation requirement appears to have been cancelled. The F7U-1s not assigned elsewhere as described above were primarily based at Vought until the time came for them to go to their final resting place in 1953.

Below, the last F7U-1, BuNo 124428, was assigned to NACA Langley in December 1952. It was powered by the production standard J34-WE-42 engines with afterburners. (NACA)

Above, BuNo 124415 on Helm's flight, like most of the last production F7U-1s deliveries, was powered by J34-WE-42 engines with afterburners like BuNo 124418 shown here. Note the modification to the boat tail to protect the end of the lengthened tailhook from the heat of the exhaust, particularly in afterburner. (Frank Gudaitis via Mark Frankel) Below, the F7U-1 fuel system was comprised of eight tanks, two in the fuselage and six in the wings. They were clustered around the center of gravity for weight and balance requirements and divided into two systems, with the secondary system not filled for interceptor-type missions where rate of climb was more important than endurance. (Author)

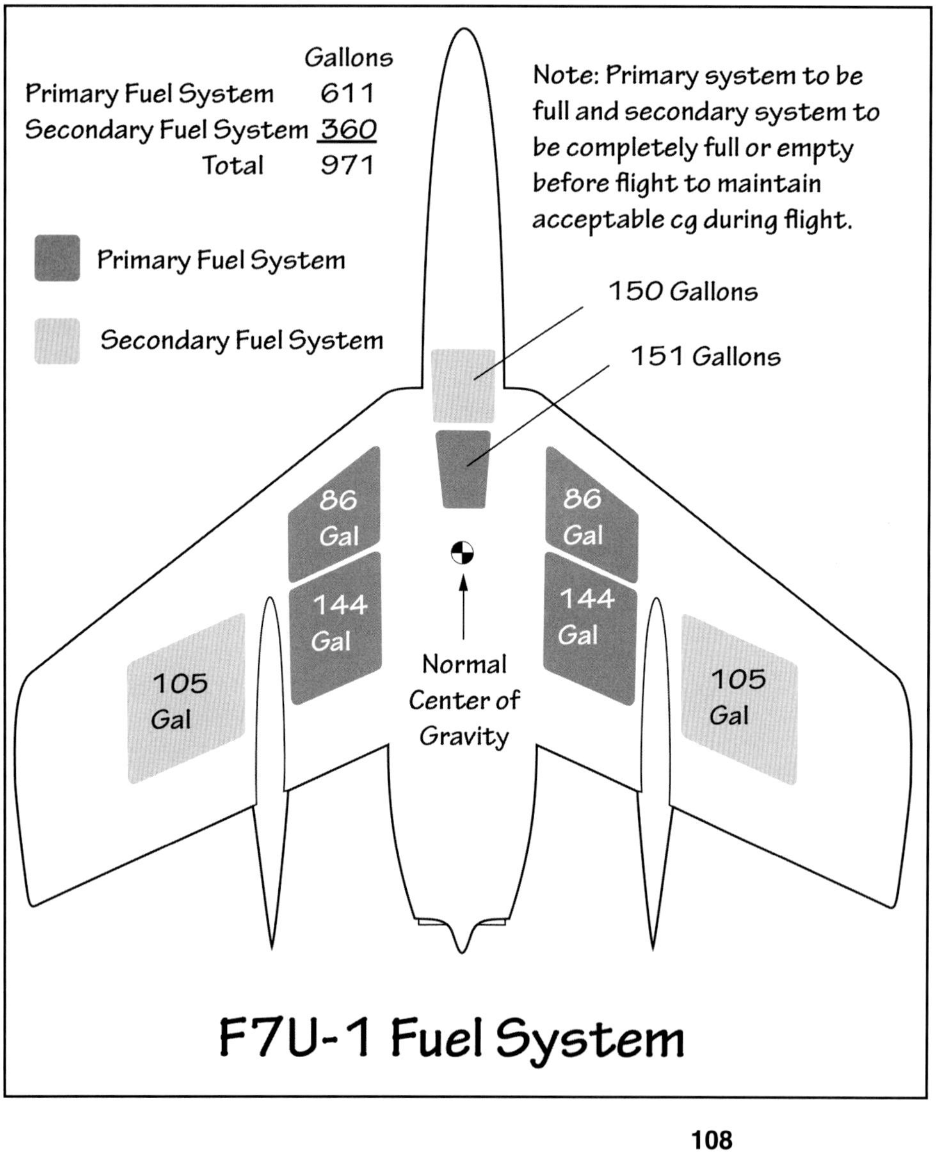

Captain Helms Stretches His Range

Jet pilots had to pay much more attention to fuel consumption than those flying propeller-driven airplanes. Jet engines had a high metabolism and consumed a prodigious amount of fuel even at idle. The J34, for example, consumed 1,140 lbs of fuel per hour at idle, 40% of the fuel burn at maximum continuous power. By comparison, the P&W R-2800 engine in the propeller-driven Corsair and Bearcat burned a little over 1,200 lbs of fuel an hour at maximum continuous power and idled at less than 180.

Compared to a propeller-driven airplane of the same gross weight, a jet therefore had about half the endurance even when loaded with twice the amount of fuel. Range was not as highly penalized, however, since long-range cruise speed at altitude was much higher than that of a propeller-driven airplane.

One of the drawbacks to a twin-engine jet, if it were capable of flying with one engine inoperative and therefore had higher installed thrust, was that it consumed more fuel than a single-engine counterpart when operating at the same level of thrust. The reason was that because of the amount of fuel required just to run a jet engine at idle, fuel burned per pound of thrust decreased significantly as thrust increased. Shutting down one engine for cruise or loiter reduced the fuel burn for the same thrust by about 50%.

The Navy therefore originally expected that the F7U would cruise with one engine shut down, with the remaining one then operating at a more efficient level of thrust, minimizing the twin's disadvantage. (Vought subsequently determined that if there was no restriction on cruise altitude due to traffic or adverse winds aloft, it was better from a range standpoint to cruise on two engines at a higher altitude than would be possible on one engine.)

The production F7U fuel system

was somewhat more complicated than originally proposed. The number of tanks went from seven to eight and the system was divided into a primary and secondary system, in part for convenience in fueling for different missions, since a deck-launched interceptor would need to be light for climb performance and theoretically didn't need as much endurance. The primary system consisted of the fuel tanks in the inboard section of the wing and the more aft of the two fuselage tanks; the secondary system was comprised of the forward fuselage tank and the tanks in the outer wing panels, which were all farther from the center of gravity. The fuel in the secondary system was used to replenish the fuel in the primary system as it was burned; like the pumps in the primary system tanks, the ones in the secondary system run on timers so that all the tanks in the secondary system emptied at the same time. However, when the afterburner was on, all of the pumps in the secondary system went on so that the fuel consumption did not drain the primary tanks dry.

Captain Lynn Helms, USMC, took advantage of single engine fuel efficiency and went one step farther while ferrying BuNo 124415 back to Patuxent River in early December 1952. It had been stored at Vought for more than a year following the at-sea trials. In late 1952, it was restored to flight status and afterburning J34-WE-42 engines were installed for the remainder of the abbreviated F7U-1 Service Trials program. The first stop of his flight was planned to be Maxwell AFB at Montgomery, Alabama, 650 miles east.

Passing Monroe, Louisiana, not quite halfway to Maxwell, Helms discovered that fuel wasn't transfering from the secondary system to the primary system. His intention to cut the flight short and land at Jackson, Mississippi, was thwarted by an unacceptably low ceiling there. Meridian, 90 miles further along his route of flight, was reporting good weather so that became his destination.

Since fuel was becoming critical, he shut down one engine on the descent toward Meridian. The bad news continued, with the weather at Meridian deteriorating to the point where a time-consuming instrument approach would be required. To maximize the amount of fuel available for a missed approach and second try (or at least a climb to altitude for ejection), he shut down the other engine and began a dead-stick instrument approach to Meridian, gliding down as he made the requisite turns on its radio range. After he broke out at 1,700 feet over the airport, Helms restarted an engine and got the Cutlass down and stopped on a 3,000-foot runway.

Raymond R. "Bud" Yinger was in Vought Flight Operations at the time: "En route (to Patuxent River, Capt Helms) developed a fuel pressure problem, so he landed in Meridian, Mississippi. Al Brill quickly prepared the Learstar for departure, with Jerry Scharfenberger (engineering), me, and two other mechanics on board. As we arrived in Meridian (on 9 December), word came that test pilot Billy Sunday had crashed on the NAS Dallas runway in an F7U-3, and had been killed. Al and Jerry returned to Dallas immediately. The F7U-1 was repaired—a faulty boost pump—and Captain Helms taxied his F7U-1 to the end of the runway, lit both afterburners—which set the adjacent woods on fire—and took off for Pax River. We returned to Dallas on commercial aircraft."

Helms reported in the Fall 2002 issue of the Naval Aviation Museum's *Foundation* magazine that as a result of this incident and the subsequent investigation, it was decreed that "no two fuel tanks would be on the same circuit breaker, future aircraft should have a 'pop-out' air turbine to provide a bare minimum of electrical and

Below, although assigned to BIS trials, NATC airplanes like this Flight Test Division F7U-1 would be occasionally flown to air shows for static display to promote interest in and support of naval aviation. (NMNA)

hydraulic power, all jet aircraft (not just test vehicles) should have ADF nav, main fuel tanks should have a minimum of 600 lbs capacity, etc."

F7U-1 Service Trials Completion

The airplane/engine performance portion of the abbreviated Service Trials evaluation were accomplished in early 1953 with F7U-1 BuNo 124415 at NATC. It was as disappointing as the carrier suitability and stability and control evaluations: "The model F7U-1 airplane as presently configured does not meet any of the performance contract guarantees." When its J34-WE-42 engines were accepted at Westinghouse, they were determined to be developing more than guaranteed thrust, albeit still slightly less than the thrust guaranteed for the -32 engines that were originally to be installed in production.

THRUST	-32 Guarantee	-42 Guarantee	415 Actual
Maximum (lbs)	4900	4650	4850
Military (lbs)	3370	3200	3320
Normal (lbs)	3020	2850	2910

The test airplane was grounded most of the time between the start of the evaluation on 6 March 1953 and its completion on 12 May 1953 because of unreliability and availability of spare parts:

Fuel boost pump	Six days
Starter	Four days
Radio and hydraulics	Two weeks
Hydraulics	Four weeks

The takeoff gross weight with full internal fuel (984 gallons) was 23,510 lbs. The actual weight empty, including mission equipment and 320 lbs of ballast to put the center of gravity at 16.5% MAC with the landing gear down, was 17,142 pounds. (The weight empty calculated without mission equipment was 15,790 lbs versus the weight empty guarantee of 13,320 lbs.) The evaluation weight was 20,660 lbs, which corresponded to the actual weight with combat fuel.

An F-86A-5 with a calibrated nose boom was used for a high-speed calibration of 415's airspeed and a TBM-3U trailing a static pressure "bomb" was used for a low-

Above, in mid-1953, BuNos 124418 and 420 became available for PR work that was actively supported by the BuAir Representative (BAR) office at Vought. (Vought Historical Center) Below, BuNos 124418 and 420 participated in an air show in Miami, Florida, in June 1953, with 420 still in its Vought test scheme. (Angelo Romano collection)

At right, LTCDR Ron Puckett, who was assigned to the Vought BAR, had owned XF2G NX91092 in the late 1940s and raced it in the National Air Races. He decided to take an F7U-1 to the1953 Dayton Air Show and compete with the Air Force for the Thompson Trophy. (Jim Sullivan collection) Below right, although LTCDR Puckett wasn't allowed to compete with the Air Force for the Thompson Trophy at Dayton, he put on a spectacular air show with F7U-1 BuNo 124420. (Frank Gudaitis via Mark Frankel)

speed calibration.

For some reason, probably because the contract guarantees were only provided at military and normal thrust, performance was not reported for combat thrust (afterburners operating).

The maximum speeds, corrected to a gross weight of 20,660 lbs except as noted, were measured to be:

Guarantee	Measured
Vmax at MRT at best altitude (knots)	
549	**474 (at sea level)**
Vmax at NRT at best altitude (knots)	
530	**423 (at sea level)**
Power-off stall* at 22,990 lbs (knots)	
100.5	**109**
Sea Level rate of climb at 22,990 lbs (fpm)	
5,280	**2,650 (sea level)**
Service Ceiling with NRT (feet)	
40,000	**28,500**
(Starting at 22,990 lbs)	

*** At the aircraft center of gravity tested, there was not enough elevator control power to stall the airplane so this was the minimum flying speed instead.**

Most of the shortfall in performance relative to the stated guarantees can be attributed to the weight increase. There is also evidence of less of a deficiency in the Air Force's 1951 evaluation. Their pilots recorded a top speed in level flight of Mach .92 at 20,000 feet, which would equate to 565 knots compared to the SAC estimate of 575 knots, not a significant shortfall, particularly since it was attained on only one afterburner.

The F7U-1's Swan Song

As more and more F7U-3s made first flights and were available to shoulder the development test requirements, the F7U-1s were increasingly redundant. Useful assignments were harder to come by for the remaining airplanes. Both 124418 and 124420, their test assignments at Vought basically complete by mid-1953, were ferried to Miami for an air show in June 1953 and 420 also participated in one in Memphis, at least as a static display, later that month.

Air shows were major events in the U.S. in the late 1940s and early 1950s. One of the biggest was the Dayton National Aircraft Show, which was held annually over the Labor Day weekend in early September. One of the events was the Thompson Trophy race for jets, which at that time was flown on a 100-km closed course. The competitors did not fly the course together but in separate timed trials.

In 1953, Vought and the Vought BAR (Bureau of Aeronautics Representative) Office decided to attend the 1953 Dayton Show and compete for the Thompson Trophy. Vought set up a six-pylon 100-km course near Dallas and Vought test pilot Jack Walton determined that the F7U-1 was competitive. Vought BAR pilot LTCDR Ron Puckett then made several practice runs around the course with the fastest circuit timed at 689 mph (599 knots), easily exceeding the speed of the 1951 Thompson Trophy race (1952's had been cancelled because of the Korean War) of 635.686 mph set by COL Fred J. Ascani in an F-86E.

LTCDR Puckett was an air race enthusiast as well as a Naval aviator. In 1947, he had purchased a surplus Goodyear F2G Corsair for $1,250 and competed in the Cleveland Air Race in 1947 and 1949. He came in second in 1949 and then sold the Corsair. He subsequently transitioned to jets and flew with VF-11 during the Korean War.

At the time, the official 100-km speed record was 652 mph set by Jacqueline Cochran in May 1953 in a

Canadair-built Sabre Mk. 3; it was even less stock than the usual speed record aircraft, since this particular airplane was the test bed for an Avro Canada Orenda jet engine that was rated 8,000 lbs of thrust, 2,000 lbs more than the F-86F's J47. However, the Air Force competition was to be an F-86D. Its J47 engine thrust had now been augmented with afterburning to about the same rating as the uprated Orenda. In July, LtCol William Barnes set an official speed record of 715.7 mph in an F-86D. However, it was accomplished by time trials on a 3-km straightaway course so speed-robbing turns were not required. Also, the course was on the Salton Sea, a desert in Southern California that is 236 feet below sea level on a very hot (104 degree) day. The temperature raised the speed of sound and thereby delayed the onset of compressibility drag with the lowest possible altitude minimizing the effect of the high temperature on engine thrust. The F-86D could not achieve the same speed on a closed course and cooler day at a higher altitude.

Puckett and CDR F.W. (Red) Brooks took two F7U-1s, BuNos 124418 and 420, to the Dayton show. The markings on these Cutlasses were similar to those of the Blue Angels, with a large "U.S. Navy" in white rather than yellow on the forward fuselage. Contemporaneous newspaper reports state that only the Air Force entry was allowed to fly for time, "for economy reasons." However, according to Vought Airport Manager Raymond (Bud) Yinger, who accompanied the F7Us to Dayton, Puckett flew the course. If he did, the Cutlass either wasn't fast enough or a time was not recorded. In any event, the Department of Defense had previously decreed that the services would not compete with each other in air show races. It would have been close, although Puckett was confident, quoted in a 3 September article in the Dayton Daily News: "The Cutlass will take the F-86 in any race of this sort. It can carry more fuel and its twin afterburners and good handling qualities make it a wonderful bird." USAF BGEN J. Stanley Holtoner's winning speed of 690.118 mph and Puckett's best of 689 mph are not directly comparable without knowing both the ambient temperature and altitude flown but it clearly would have been close.

Below, when the F7U development became protracted, the Navy was desperate to deploy a swept-wing carrier-based jet fighter. The Bureau of Aeronautics contracted with North American to marinize its F-86 as the FJ-2 Fury. Here, an early production FJ-2 is on static display alongside F7U-1 BuNo 124418 at the 1953 Dayton air show. (Frank Gudaitis via Mark Frankel)

Nevertheless, the Vought BAR pilots put on quite a show. LTCDR Puckett, who had flown with VF-11 in Korea and raced civil Corsairs, was the F7U demonstration pilot at the 1953 Dayton Air Exposition. As part of his performance, he made an inverted pass at 500 miles per hour, reportedly at only 200 feet off the ground. He also demonstrated the dual-position main landing gear when he taxied by the crowd while rocking up and down. Following the air show, he ferried BuNo 124420 to the Naval Air Technical Training Center at NAS Jacksonville where it was to be used for ground training of enlisted mechanics, electricians, and ordnance men. BuNo 124418 arrived at NAS Memphis on 11 September to join the former Blue Angel Cutlasses and 124424 already there as maintenance trainers.

CDR Hill's Wild Ride

On 23 October 1953, CDR John S. Hill took off from NAS Patuxent River in F7U-1 BuNo 124419 to ferry it 93 miles south to NAS Norfolk, Virginia. It was then to be loaded aboard *Coral Sea* for "noise level tests" according to the accident report. It's not clear what this was all about although presumably, this was related to the operation of 419's afterburners for launch (415 did not have afterburners during its at-sea trials). Coral Sea was hosting a number of different NATC aircraft at-sea evaluations at the time, including the F7U-3, XF4D, and XF3H, among others. It's doubtful that any landings would have been risked, but some run-up's in afterburner in front of the blast deflectors behind the catapults and a launch for its return to Pax may have been its planned usage.

Shortly after takeoff, Hill had a hydraulic flight control system boost failure. He shifted to manual mode and continued on. As he neared NAS Norfolk, he attempted without success to reengage the hydraulic flight control system. In accordance with the flight manual emergency procedure, he then conducted a slow flight evaluation on manual control and determined that the minimum control speed was 130 knots. He therefore elected to make the approach at 140 knots. Although strictly speaking an emergency, the F7U-1 had been landed many times in manual mode without incident, both to evaluate the capability and because of the system's lack of reliability.

CDR Hill was a very experienced pilot, with almost 5,000 total flight hours, but he had only 10.6 hours in the F7U-1. On short final, probably because he was concerned about being fast on approach, he extended the speed brakes. By design, they were very effective at killing speed; the result was a sudden and significant loss of lift and corresponding increase in rate of sink. His application of aft stick and forward throttles were ineffective at cushioning the landing under the circumstances. The Cutlass hit so hard, nose wheel first, that it bounced 50 feet back into the air. Because at least one motion picture photographer was in place to document the arrival of the Navy's newest fighters at Norfolk for their at-sea trials, Hill's arrival and subsequent excursion off the runway was captured on film.

The sink rate on that first touchdown was estimated to be more than 50 feet per second, three times the design requirement. If that estimate was accurate, Vought engineers had clearly been very conservative with their structural analysis. Instead of the landing gear collapsing, the only

Below, For some reason, BuNo 124419 was the only F7U-1 to have a distinctive paint scheme on the vertical fins. The checkerboard was originally only on the outboard side of the fins but at some point prior to transfer to NATC, possibly after the structural modification to the vertical fins, it was painted on both sides. (National Naval Aviation Museum)

obvious failure observed was the blowout of the nose wheel tire. However, it seems likely that it was then that the forward fuselage structure was damaged almost to point of failure and CDR Hill suffered three chipped vertebrae in his back.

Almost immediately after touching down the second time, the airplane careened off the runway into the grass field to the left, probably due to the blown nose wheel tire and/or Hill being momentarily stunned. After 200 feet or so, the nose landing gear broke off followed by the forward fuselage hitting the ground and then breaking off to the left, although still connected to the rest of the fuselage by control cables, electrical harnesses, and some structure. Shortly thereafter, with the airplane still traveling at about 80 knots, its right wing hit an asphalt tank truck parked off to the side of the runway and was torn off, precipitating a spectacular fireball.

The fuselage, with the left wing still attached, was spun around 180 degrees by the encounter with the truck and came to a halt. During this gyration, the forward fuselage containing CDR Hill switched sides so it was now to the right of the rest of the main wreckage. Fortunately, Hill was not further injured by his wild ride and the wind direction was such to keep the fire from enveloping the cockpit. He was able to unstrap, crawl out, and—with the help of a workman at the scene of the crash—walk away as the fire fighting equipment arrived.

Unfortunately, Frank Harris, one of the four workmen injured during the crash, died of his burns.

Above, the cockpit wound up lying on its left side, facing the opposite direction from the fuselage. Not only did Hill survive, he was seen having a drink at the NAS Norfolk BOQ that evening. (NA 80-G-629993) Below, While making a boost-off landing at Norfolk, Virginia, CDR Hill got too slow on short final and developed a high sink rate, with disastrous results. (Author from 16mm movie film stills via Wayne Weiss)

Cdr. Hill's Wide Ride

Due his high sink rate on short final, Cdr Hill bounces 50' back into the air and goes off the runway to his left, which collapses the nose landing gear.

The forward fuselage breaks off, fortunately to the left, because the Cutlass subsequently hits an asphalt tank truck with the right wing.

The cause of the hydraulic failure was determined to be a loss of air pressure in the hydraulic accumulators after takeoff due to a component failure that was not established.

Like Paul Thayer, Hill had survived a spectacular crash in an F7U-1 without significant injury. The accident report noted that: "The structure of the cockpit and the attachments therein are well designed to protect the pilot and Change Vought should be commended for their considerations of those safety factors in the design of this plane."

Epilogue

As a result of placing emphasis on general-purpose jet fighters and forgoing opportunities to either accept the F7U-1 with deficiencies or fund proposed alternatives such as a swept-wing F6U or F2H, the Navy found itself without a high-performance day fighter when the Korean War demanded one. It recovered quickly, contracting with North American to modify the F-86 for carrier operations and Grumman to put swept wings on the Panther, creating the F9F-6 Cougar. As a result, the latter was available for deployment in 1953. (The swept-wing Fury turned out to require a more powerful engine for carrier duty so the FJ-2 was assigned to shore-based Marine Corps squadrons).

Most of the Air Force's swept-wing and delta-wing high-performance fighters that followed the F-86 didn't fare much better than the Cutlass until the so-called Century Series programs. The McDonnell F-88 program floundered like the F7U-1 did for somewhat the same reasons. It morphed into the bigger F-101 much like the F7U-1 became the F7U-3. Kelly Johnson admitted that the F-90 was an overweight failure. The Republic F-91 demonstrated that auxiliary rockets weren't a good solution for lack of jet engine thrust. The delta-wing XF-92 was relegated to a research role. The F-93 and F-95 were variations on the F-86; the former didn't reach production and the latter was redesignated the F-86D

Above, after ignoring the F-86 as a candidate for a carrier-based day fighter for five years, BuAer reversed course and provided North American a contract to marinize it. The performance of the resulting FJ-2 was disappointing so it was given to the Marine Corps and North American installed a more powerful engine to create the FJ-3 (NA 80-G-708996) Below, Grumman also got a contract to sweep the wings on the F9F-5 Panther. The result, the F9F-6, was flying in only six months and deployable two years later. (NA 80-G-708517) Bottom, from the standpoint of developing a swept-wing jet fighter with afterburners, the Air Force didn't have much more luck than the Navy with either its F-88 (see page 66) or the F-90 shown here. Neither went into production. (USAF)

and did. Like the Navy's F9F-6, the Air Force F-96 was a swept-wing variant of the F-84; it was successful and was redesignated F-84F. It wasn't until the North American F-100 that the Air Force was provided with a truly supersonic replacement for the F-86 and even it had teething problems that took time and lives to resolve.

Even if the Navy had stayed the course and continued with production of the -1 or even the -2, Vought and Westinghouse would have needed to come up with significant improvements in the Cutlass hydraulic system, maintenance access, and engines. The production -1s were notorious for hydraulic problems. The two that were assigned to the Blue Angels were reportedly parked over big trays that collected the leakage from the hydraulic system. The one bailed to Convair is remembered as being a "massive hydrulic leak." In its service trials, BuNo 124415 suffered a major hydraulic component failure every week. F7U-1 maintainability was so bad that one of the F7U-3 design goals was improvement of maintenance access. The afterburner was an ongoing problem from an availability and operational standpoint. And even if that had been fixed, the F7U-1's empty weight eventually grew to almost 16,000 lbs, meaning even more thrust had to be wrung out

Above, all three of the XF7U-1s crashed as did three of the 14 F7U-1s. Probably the longest-lived survivor was the first production F7U-1, BuNo 124415. It was on display at the Navy maintenance school at NAS Memphis through at least August 1957. Unfortunately, it was subsequently scrapped along with all the other F7U-1s that survived flight test. (John McKenna) Below, the airframe of an F7U-1 is barely recognizable under some crude modifications to make it resemble another airplane for which a Davis barrier or barricade engagement evaluation was necessary. (Author collection)

At left, a formation of the newest 1954 Navy fighters assigned to VC-3. The F2H-3 in the slot represented its primary mission of providing detachments of night fighters to deployed air groups. One of its collateral duties in 1954 was as the west coast Transitional Training Unit for newly assigned Commanding and Executive Officers of high-performance jet-fighter squadrons who were not current. The types assigned were the FJ-3 leading the formation, F9F-6, and F7U-3. Although the basic Cutlass design predated the other two swept wing fighters by several years, the F7U was only now being readied for deployment as they were. They would also be operational several years longer than it would be. (NMNA)

of the J34 itself for the early Cutlass to have acceptable performance. It therefore seems unlikely that there was any chance of combat-worthy J34-powered Cutlasses being deployable in time to take on MiGs in the sky over Korea.

Although the F7U-3 eventually deployed, it proved to be a disappointment. In order to provide acceptable reliability and durability, the Westinghouse J46 engine had been derated. The reduction in thrust combined with the weight growth during its development meant that the F7U-3 performance did not meet expectations. (On its final cruise however, from an angled-deck carrier, it successfully met all its operational assignments, deploying with the Sparrow I missile.)

And if nothing else, the disappointing F6U and F7U programs provided Vought and the Navy with the experience that resulted in the design of one that exceeded expectations, the F8U Crusader.

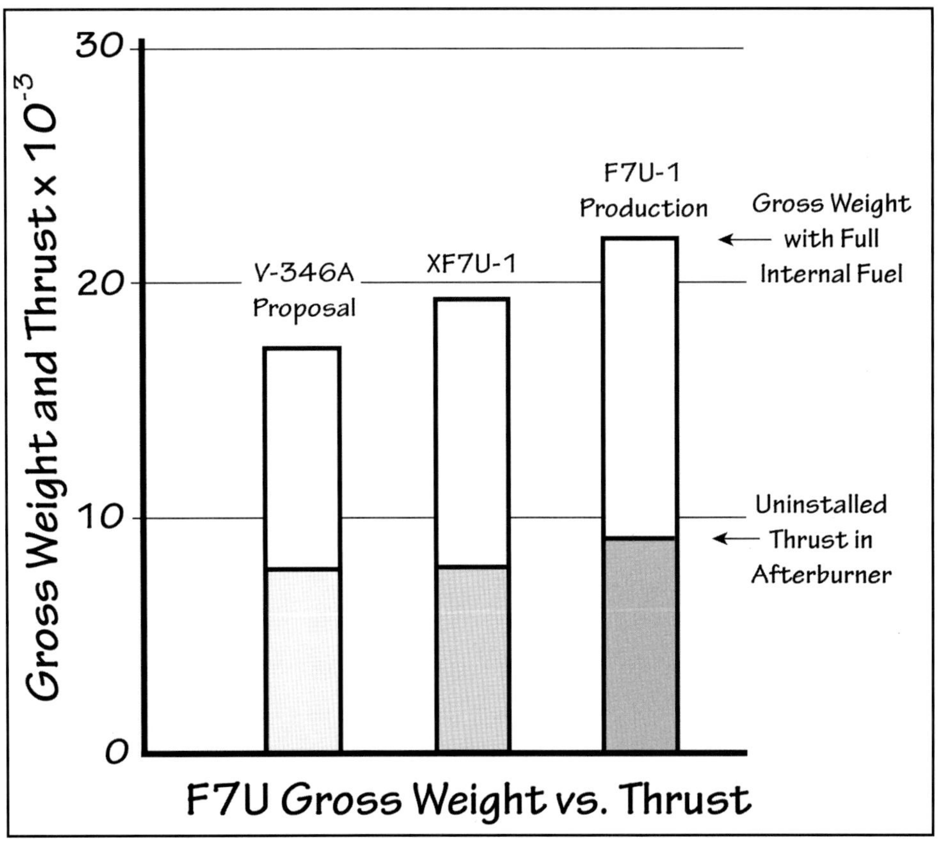

At top right, in the search for its first supersonic fighter the Air Force tested the North American swept wing YF-93A as an interim aircraft between the F-86 and F-100. (USAF) Above right. the supersonic YF-100 in flight. (USAF) At right, because of the significant underestimate of weight and only limited growth in the Westinghouse J34 engine thrust, among other things, it's likely that the F7U-1/2 would have disappointed even if the Vought had been able to complete its development and qualification to the original plan. (Author)Below, in the end, however, the experience that Vought and the Navy gained from two disappointing jet fighter programs yielded an airplane, the F8U Crusader, that they could be extremely proud of. (Vought Heritage Center)

Final Disposition

XF7U-1

122472 Crashed 28 September 1949 at Ardmore, Oklahoma

122473 Crashed 14 March 1949 at NAS Patuxent River, Maryland

122474 Crashed at 7 July 1950 NAS Patuxent River, Maryland

F7U-1

124415 Last known assignment, NAS Memphis, Tennessee, 24 July 1953

124416 Crashed 6 July 1950 at NAS Dallas, Texas

124417 Crashed 28 September 1950 at NAS Dallas, Texas

124418 NAS Memphis NATTC, retired 11 September 1953 with 68 hours total time, stricken Nov 53

124419 Crashed 23 October 1953 at NAS Norfolk, Virginia, and destroyed with 123 hours total time.

124420 NAS Jacksonville NATTC, retired 14 September 1953 with 145 hours total time, stricken Nov 53

124421 NAS Point Mugu, retired 16 September 1953 with 44 hours total time, stricken Nov 54

124422 NAS San Diego, retired 11 November 1954 with 134 hours total time, stricken June 55 at NATF Lakehurst

124423 NATC Patuxent River, retired 20 August 1953 with 65 hours total time, stricken Aug 53

124424 NAS Memphis, retired 14 July 1953 with 43 hours total time, stricken Oct 53

124425 NAS Philadelphia, retired 12 October 1953 with 56 hours total time, stricken Jan 1954

124426 NAS Memphis NATTC, retired 4 June 1953 with 26 hours total time, stricken Sep 53

124427 NAS Memphis NATTC, retired 4 June 1953 with 26 hours total time, stricken Sep 53

124428 NAS Philadelphia, retired 16 March 1954 with 22 hours total time, stricken June 1954

Note: Total flight times and final location from the Aircraft History Cards.

Below, Vought's first two jets shared a common failing: not enough thrust from the basic engine(s) and too high a fuel consumption when in afterburner. (Vought Heritage Center)

F-86 and F7U Program Milestones

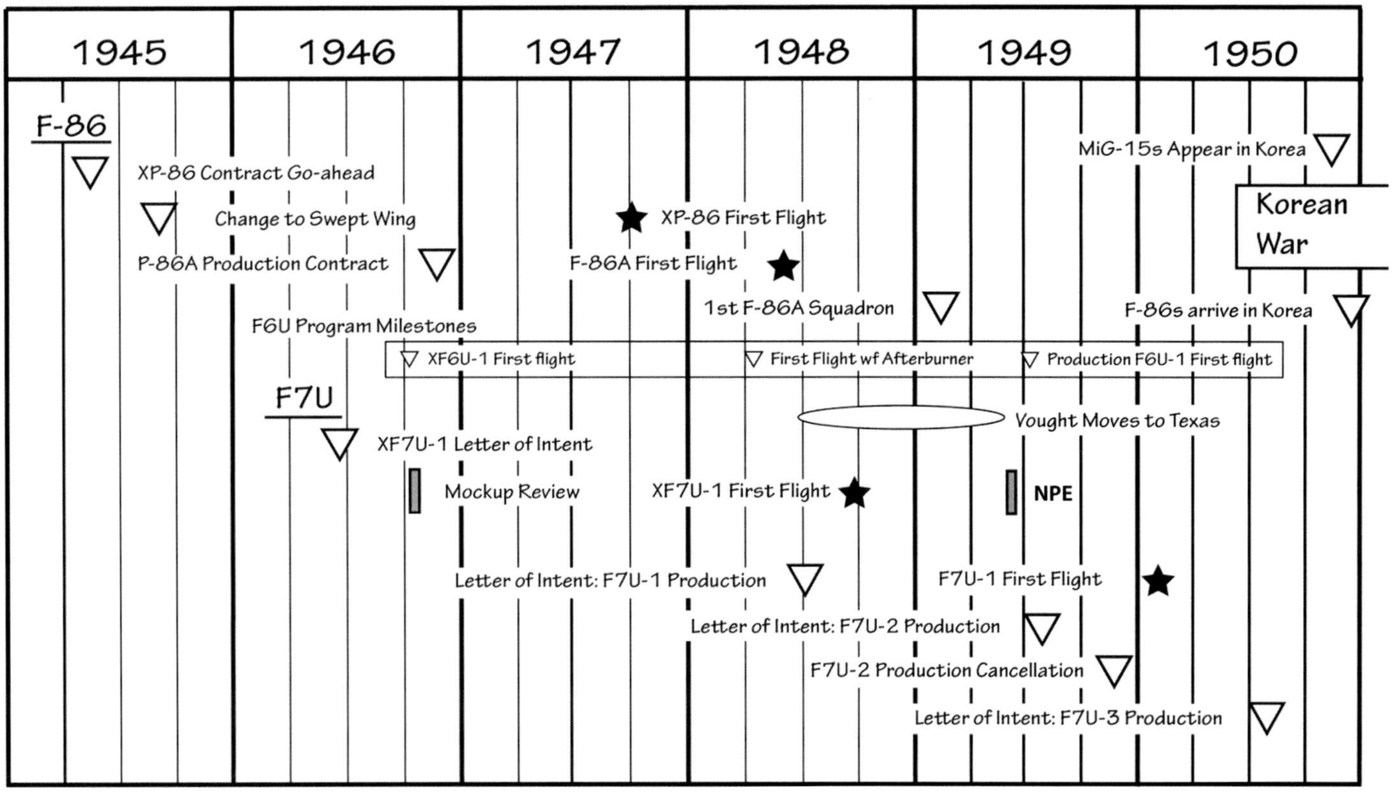

INDEX NO.	NOMENCLATURE	FIGURE NO.
	GENERAL ARRANGEMENT	
	WING ERECTION	4
1	Speed Brake Assembly	5
2	Slat Installation — Center Section	7
3	Slat Installation — Outer Panel	8
4	Ailavator Assembly	10
5	Outer Panel Assembly	14
6	Tip Installation — Outer Panel	15
	INBOARD PROFILE	29
	Front Fuselage Section	29
7	Nose Cone Installation	33
8	Windshield and Cowl Installation	34
9	Sliding Section Installation — Cockpit	36
10	Door Installation — Nose Gear	42
11	Armament Installation	50
12	Seat Installation — Ejection	52
13	Nose Landing Gear Installation	67

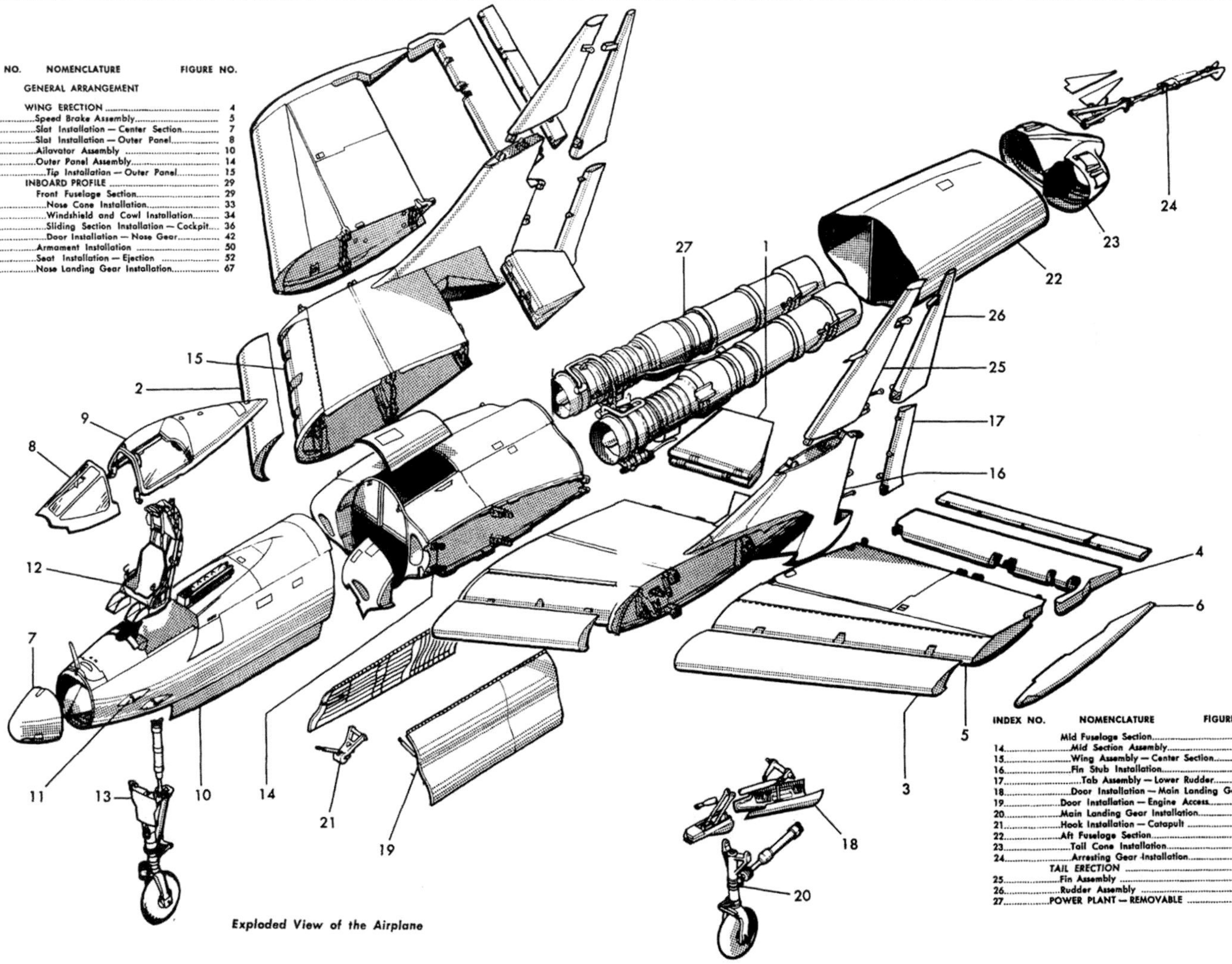

INDEX NO.	NOMENCLATURE	FIGURE NO.
	Mid Fuselage Section	75
14	Mid Section Assembly	75
15	Wing Assembly — Center Section	84
16	Fin Stub Installation	88
17	Tab Assembly — Lower Rudder	89
18	Door Installation — Main Landing Gear	92
19	Door Installation — Engine Access	101
20	Main Landing Gear Installation	103
21	Hook Installation — Catapult	105
22	Aft Fuselage Section	107
23	Tail Cone Installation	107
24	Arresting Gear Installation	110
	TAIL ERECTION	125
25	Fin Assembly	126
26	Rudder Assembly	127
27	POWER PLANT — REMOVABLE	129

Exploded View of the Airplane

F7U-1 CUTLASS COCKPIT LAYOUT

The production F7U-1 cockpit as depicted in the flight manual

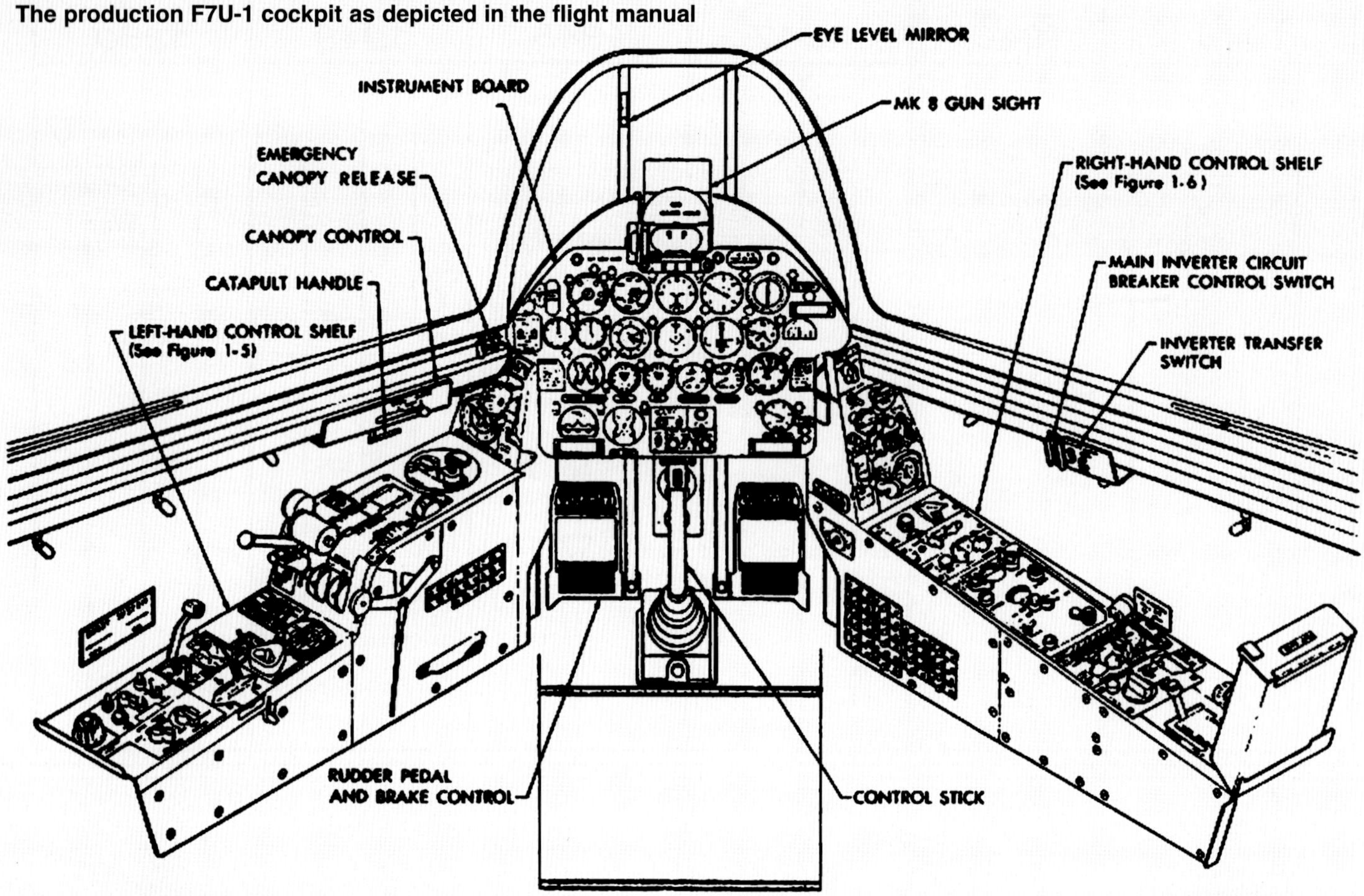

F7U-1 LEFT-HAND CONSOLE

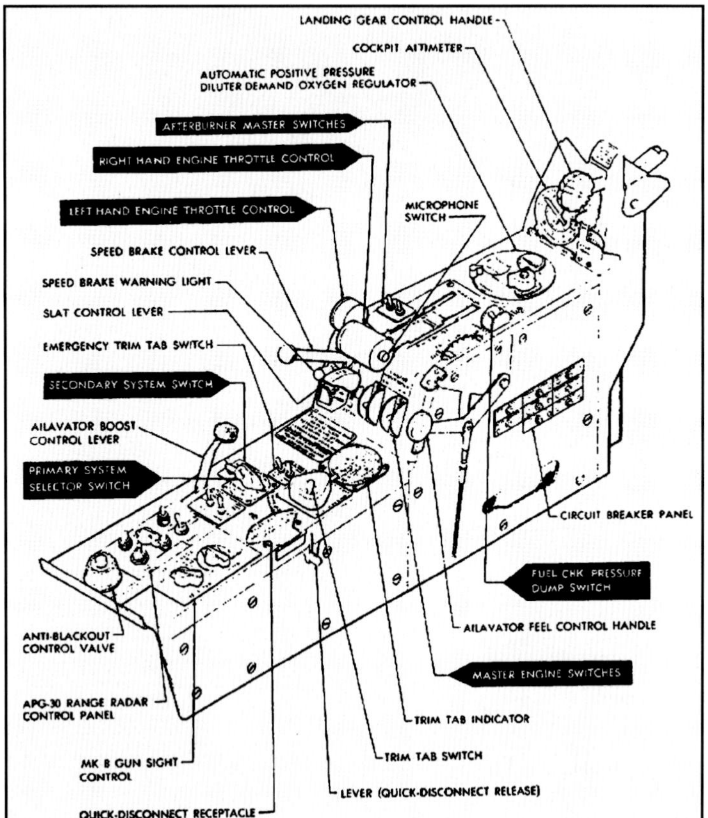

F7U-1 RIGHT-HAND CONSOLE

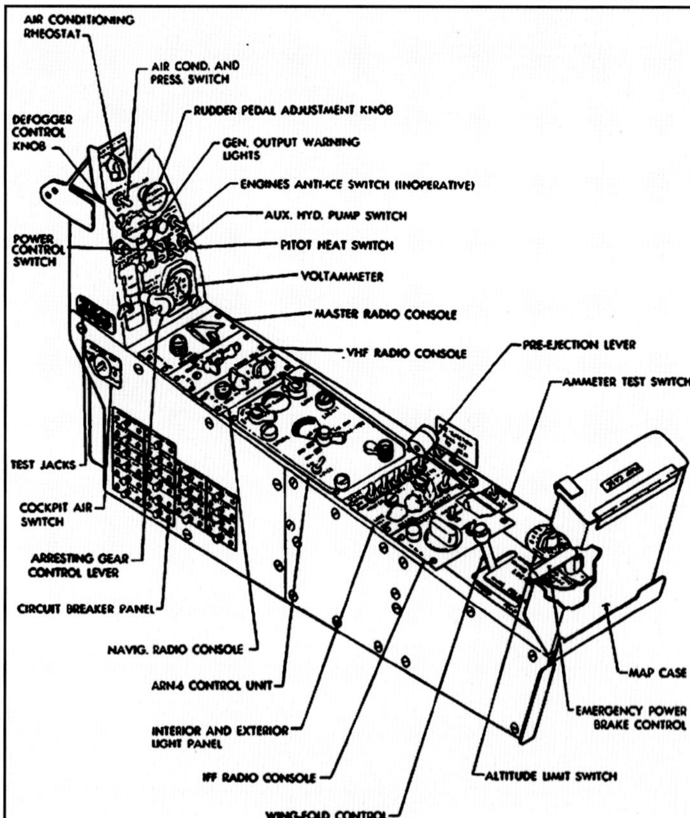

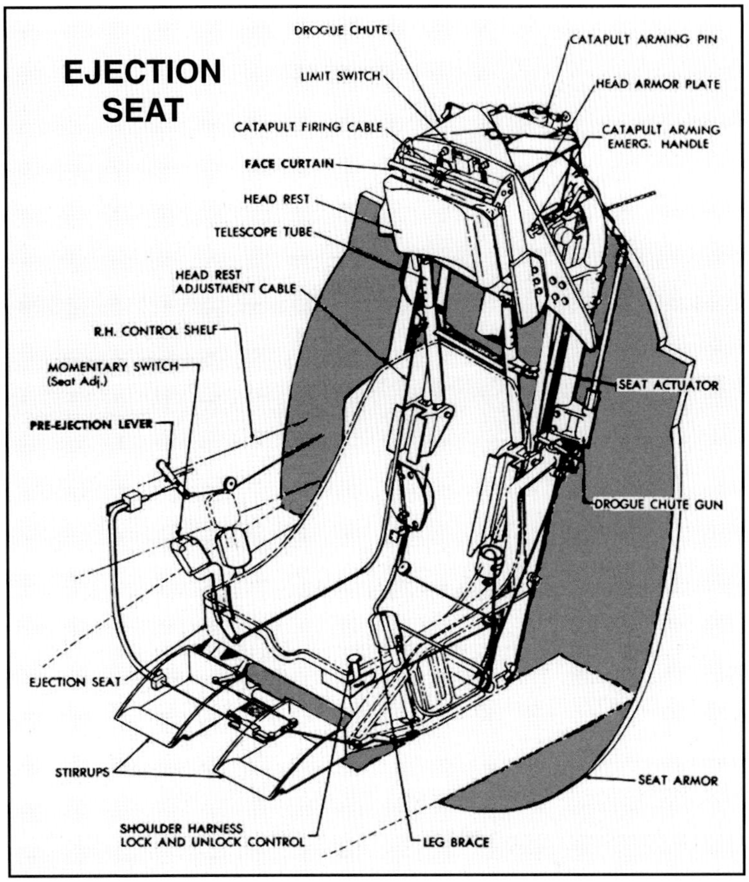

The XF7U-1 cockpit (above) was configured to incorporate the large and removable chart board that slid out between the upper and lower sections of the instrument panel. Flight-test-only switch panels and instruments were scattered about the cockpit. At right, the early Navy ejection seats had to be armed by pulling the pre-ejection lever on the right side of the cockpit. This lowered the seat, raised the leg braces, locked the shoulder harness, decompressed the cockpit, and jettisoned the canopy. A red light on the instrument panel then indicated that all was in readiness for the pilot to pull the face curtain and be ejected

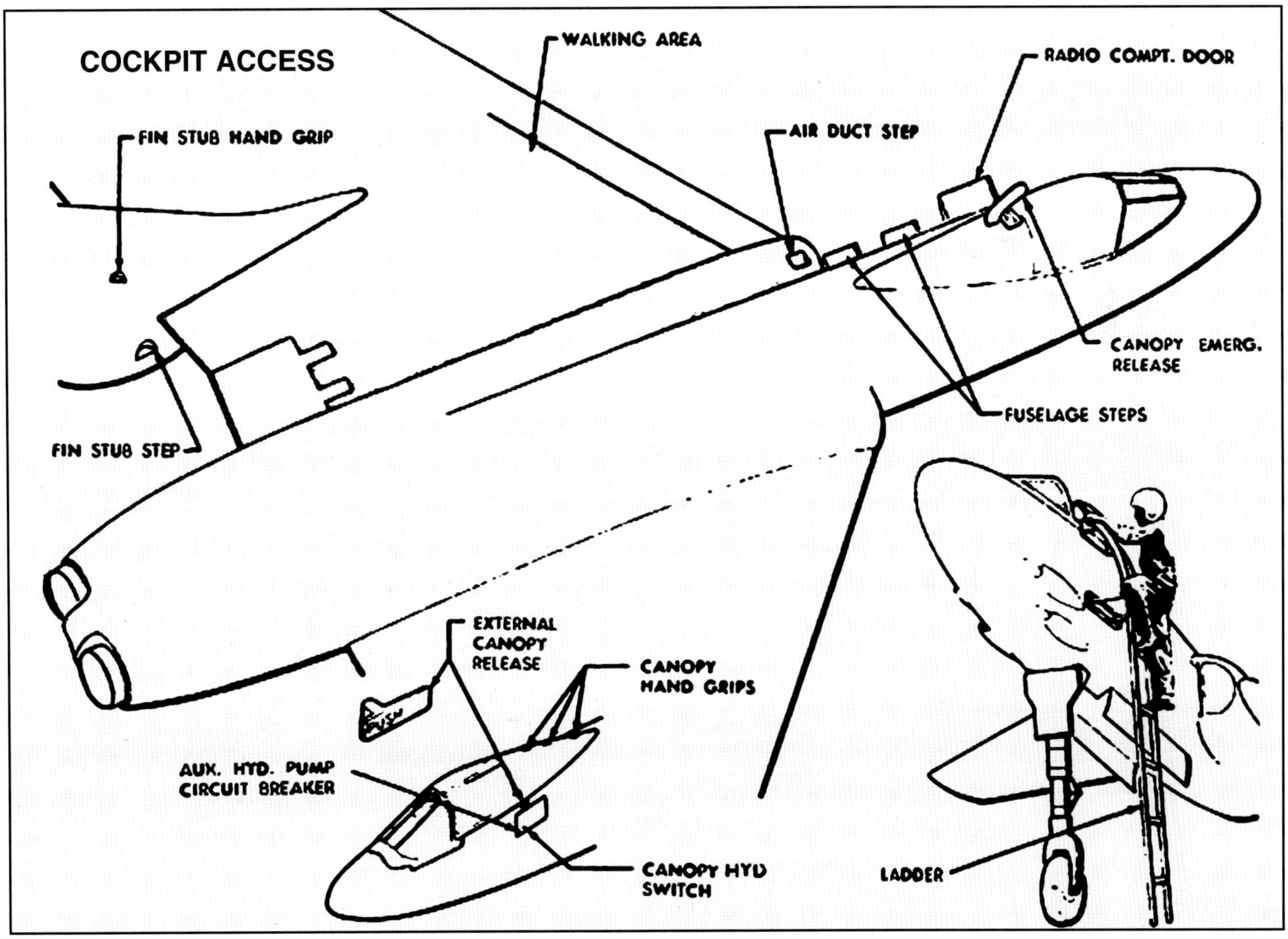

Above, the introduction of the steam catapult ended concerns about maximizing the F7U's angle of attack during its launch from the carrier. The end speed that could be achieved even at the F7U-3M's takeoff weight provided more than enough lift. (NA 80-G-102778) Below, the production F7U-1 looked very much like the proposed V-346A but under the skin virtually everything had changed in detail from the radar-ranging gunfire control system to the tailhook. In the process, the empty weight had increased by 6,500 lbs, about 70%, without any increase in wing area or much in the way of engine thrust. Like many attempts at a great leap forward, this one fell short of its goal.

F7U-1 CUTLASS CUTAWAY

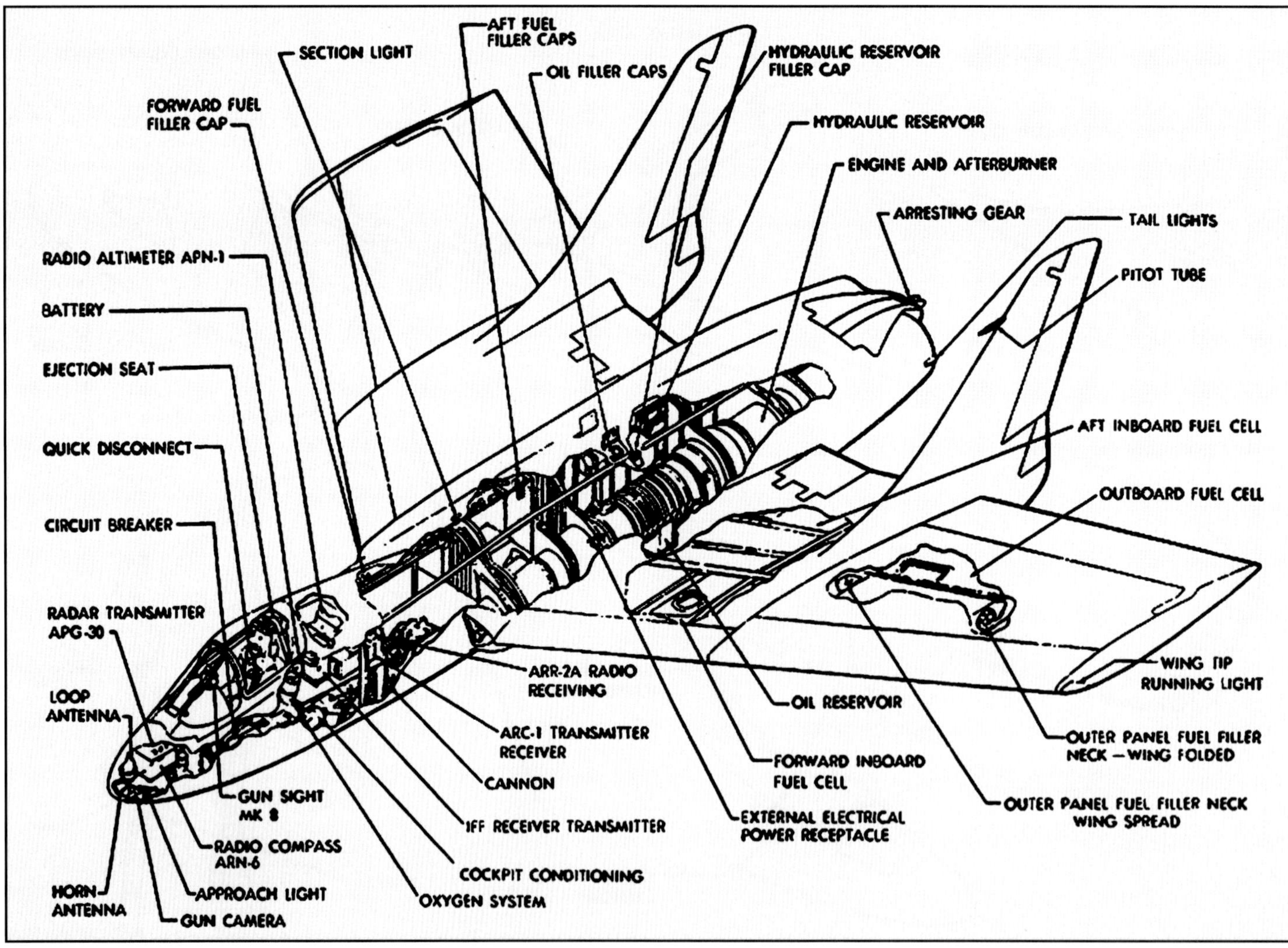

XF7U-1 BuNo 122472
Red nose and boom with yellow diamond outlined in blue or black

XF7U-1 BuNo 122473
White(?) nose, boom, and "Cutlass" with orange(?) "473" and red(?) "Chance Vought"

XF7U-1 BuNo 122474
Dark yellow or light orange nose and boom with red diamond outlined in blue or black

F7U-1 BuNo 124416
Red or maroon nose and boom and unknown "badge" with a yellow top half and white lower front quarter

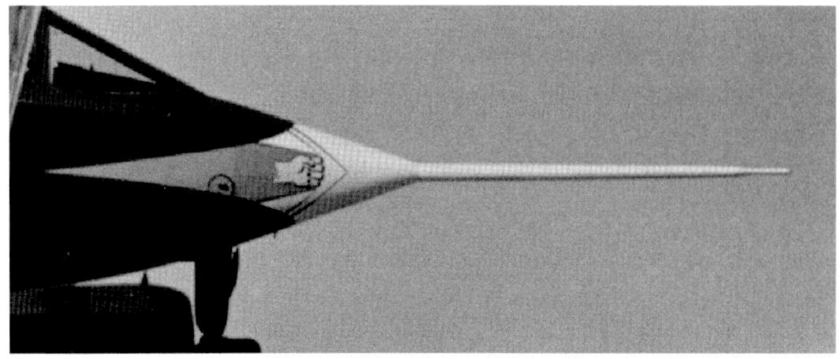

F7U-1 BuNo 124418
White nose , boom, and fist with an orange panel behind and aft of the fist

F7U-1 BuNo 124419
White nose and boom with a red arrow

F7U-1 BuNo 124420
White nose and boom with a gold(?) zigzag outlined in black or blue

F7U-1 BuNo 124423
White(?) nose and boom with a red(?) arrow

Vought Flight Test XF7U-1 and F7U-1 Nose Markings

1/86 SCALE TRIM / LINDBERG XF7U-1 CUTLASS KIT

Trim Model Products produced a small line of injection-molded kits in the early 1950s, one of which was an F7U-1. Lindberg acquired the molds in the early 1960s and produced it, first individually and then in a boxed set. The kit represents the XF7U-1 with the original speed brakes and the later, taller vertical fins. The shape issues are noticeable: the vertical fins too narrow in chord, a nose that is circular in cross section rather than oval, and a canopy that only approximates the shape of the actual one particularly with respect to the relative size of the windscreen. For some reason, the tip of the aft fuselage between the engine exhausts was omitted.

One feature, albeit bogus, was the inclusion of six rockets that were to be mounted on the outboard wing panels. Not very subtle raised lines were used to denote the edges of panels, the control surfaces, the main wing spar, the vents in the top of the aft fuselage, etc. The cockpit consisted of the pilot's head and headrest molded on each of the fuselage halves. A fairly detailed landing gear and separate wheels were provided, although the fuselage mounting point of the nose landing gear is too far aft.

1/48 SCALE COLLECT-AIRE F7U-1 CUTLASS KIT

Unfortunately out of production, this is a very detailed and complete resin kit of the F7U-1. It included decals by MPD for a Blue Angels airplane and one assigned to Point Mugu. The ailavators and slats could be positioned. The canopy was vacuformed (and two were provided) but the landing gear and ejection seat were metal. The master was reportedly exquisite but unfortunately the resin casting and canopy forming quality varied from kit to kit, with some parts requiring replacement (frequently the canopies) or repair.

Steve Ginter built this model from the 1/86 Lindberg kit in the early 1960s in accordance with the kit's directions and using the decals provided. (Ginter) Below, The Collect-Aire kit can be built into an excellent model with some effort, as was done here by Randy Smith. (Randy Smith)

This kit, long out of production, consisted of partially shaped wooden (pine) parts that needed substantial sanding to achieve the correct shape. A clear injection-molded canopy and a full set of markings were provided but no landing gear or cockpit. The model was to be displayed on a stand, which consisted of a wooden base and a wire support.

The kit included sandpaper in two grades along with a sanding stick and a packet of casein glue powder.

The shape of the wing and vertical fins are consistent with the production configuration with the exception of the lower trailing edge of the vertical fin, which appears to have been unduly influenced by the F7U-3's and extends too far aft, and the canopy, which is a little too "peaked" in profile, the parts were accurate in basic shape and size.

The kit includes decals to represent the red interior of the slat wells since its designer didn't realize that this area was only exposed when the slats were extended.

1/72 SCALE STROMBECKER F7U-1 CUTLASS KIT

Above, All parts shaped: ready to assemble was a bit of an overstatement. Some work was required to achieve the final shape shown on this display. Note the price. The decals included white wing walk markings that weren't present on any F7U-1 and wide red lines that were only appropriate if the slats were extended. (Steve Remington) Below, In addition to removing all the raised detail, Paul Boyer modified the 1/72 Aurora kit by adding a cockpit and vacuform canopy he created. The decals are by Microscale from their Blue Angel Bearcat and Panther sheets. (Paul Boyer) Bottom, Larry Schmidt modified the 1/72 Aurora kit to add wheel wells and inlet ducts and reshape the nose. He also sanded off all the raised detail, scribed the control surface lines and added a cockpit. The canopy and landing gear are out of the box. (Larry Schmidt)

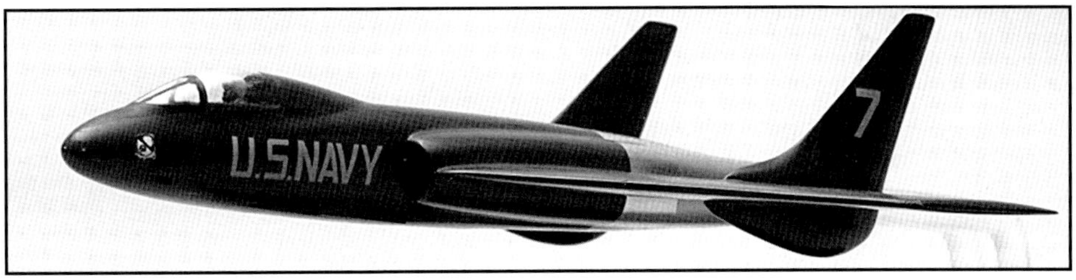

1/72 SCALE AURORA F7U-1 CUTLASS KIT

First released in the late 1959 and long out of production, the injection-molded Aurora kit can still be found. Although the illustration on the box is of an F7U-3M with a refueling probe and a weapons pylon on the outboard wing panel, the parts inside are representative of an early F7U-1 with the original XF7U-1 speed brakes and the production vertical fins.

The kit appears to be based on the original 1/86 Trim kit, only scaled up to approximately 1/72. It therefore has the same accuracy issues. Generally reported to be 1/70 scale, the fuselage is very close to 1/72 in size and shape. The wings are slightly undersized for 1/72and the vertical fins definitely so. The major drawback in accuracy is the canopy, which has a windscreen that is little more than half the size that it should be.

The kit is crude by today's standards with few parts. The representation of the cockpit consists of the pilot's head and a headrest on a flat surface at the level of the canopy rails. The parts are covered with too many rivets to count and raised lines represent all panel lines and control surface locations. Each landing gear strut and wheel is molded as one piece. The mounting point for the nose gear strut is too far aft on the fuselage, which mostly makes up for it being too short. No interior is provided for the engine inlets and there are no wheel wells.

My example was possibly never built because the one-piece wings were not given enough time to cool prior to being removed from the mold. The hot plastic at the thickest part of the one-piece wing, the inboard section, subsequently contracted unconstrained by the mold, distorting it. Mine appears to not be an isolated example, so buyer beware.

This is a resin kit produced using Vought drawings in the past few years. Unlike the Aurora and StromBecKer kits, it includes a complete cockpit. Vertical fin and wing speed brakes are provided for both the XF7U and production configurations. However, the short vertical fin isn't quite short enough and lacks the large aerodynamic balance at the top; the bottom of the rudder is also misrepresented because the lower portion of the fin is common to both configurations. The nose gear is the production configuration but both the fins and the nose gear could easily be modified to represent a prototype.

The cockpit is adequate to the scale but a bulkhead needs to be added behind the headrest to close off the canopy aft of the clear portion. The engine inlet lacks the boundary layer splitter plate not to mention an interior wall.

The vacuform canopy is a little too "peaked" in profile but acceptable. Decals are provided for both a Blue Angels airplane and F7U-1 BuNo "122422", which is unfortunately incorrect; it should be 124422. No decals are provided for an XF7U-1 other than the national insignia.

Above right, the 1/72nd Anigrand kit is easily the most accurate and buildable F7U-1 kit in this scale. (Joe Hegedus)

Revell began their production of injection-molded airplane kits in 1953. The first three were the Lockheed F-94C, Grumman F9F-6, and the Vought F7U-1. A display stand was provided in lieu of landing gear. They were subsequently released as a boxed gift set. The F7U was re-released as a -3 although the modifications were simply to add landing gear and rivet detail. A subsequent release added four missile pylons and missiles.

At right, the 1/59th Revell kits rareness exceeds its accuracy. (Al Bussie)

1/72 SCALE ANIGRAND F7U-1 CUTLASS KIT

1/59 SCALE REVELL F7U-1 CUTLASS KIT

In the early 1960s, Lindberg apparently used the 1/86 kit acquired from Trim as the basis for a larger, more popular 1/48 version.

The kit parts are very close to 1/48 but shape and other errors are present as on the original Trim kit: the nose landing gear mounting point is too far aft, the vertical fin is too small, there is no after body between the engine exhausts or inlet interior, the canopy is incorrect, the tip of the nose is round instead of a squared oval, etc. Although wheel wells are not provided, it does have a cockpit of sorts. The seat consists of a circular headrest and a perch on which to mount the pilot (who appears to be wearing a leather helmet and crossed bandoliers although I suppose they are intended to represent a shoulder harness configuration with which I am not familiar): there are no parts for the instrument panel, consoles, controls, etc.

This injection-molded kit was recently reissued (without the stand provided with the original or the nose boom) and so is readily available. The obtrusive and unnecessary rivet detail on the original parts has been significantly reduced with the reissue. Panel and control surface demarcation is by finely raised lines. The wheels and landing gear struts are separate.

Like the other injection-molded kits, it is acceptable as a manufacturer's desktop model of the F7U-1 or the starting point for a major conversion to a more accurate model of the original Cutlass.

1/48 SCALE LINDBERG F7U-1 CUTLASS KIT

Above, the many reissues of the 1950s Lindberg kit provides a very rough basis for a 1/48 F7U-1. This box top features a photo of an out-of-the-box built model that has a more pointed nose than the one in the box, possibly because it was built from the original kit that had the research nose boom. The decals are on an odd mixture of out-of-date (no red bars on the national insignia) and before its time (red inlet and yellow rescue markings). (Author) Below, Lindberg F7U-1 built straight from the box in 1980 with assorted aftermarket decals used to depict BuNo 124419. (Ginter) Bottom, Lindberg F7U-1 built in 2012 as Blue Angels Seven is compared with the Hobbycraft F7U-3M Cutlass. (Mike Castro)

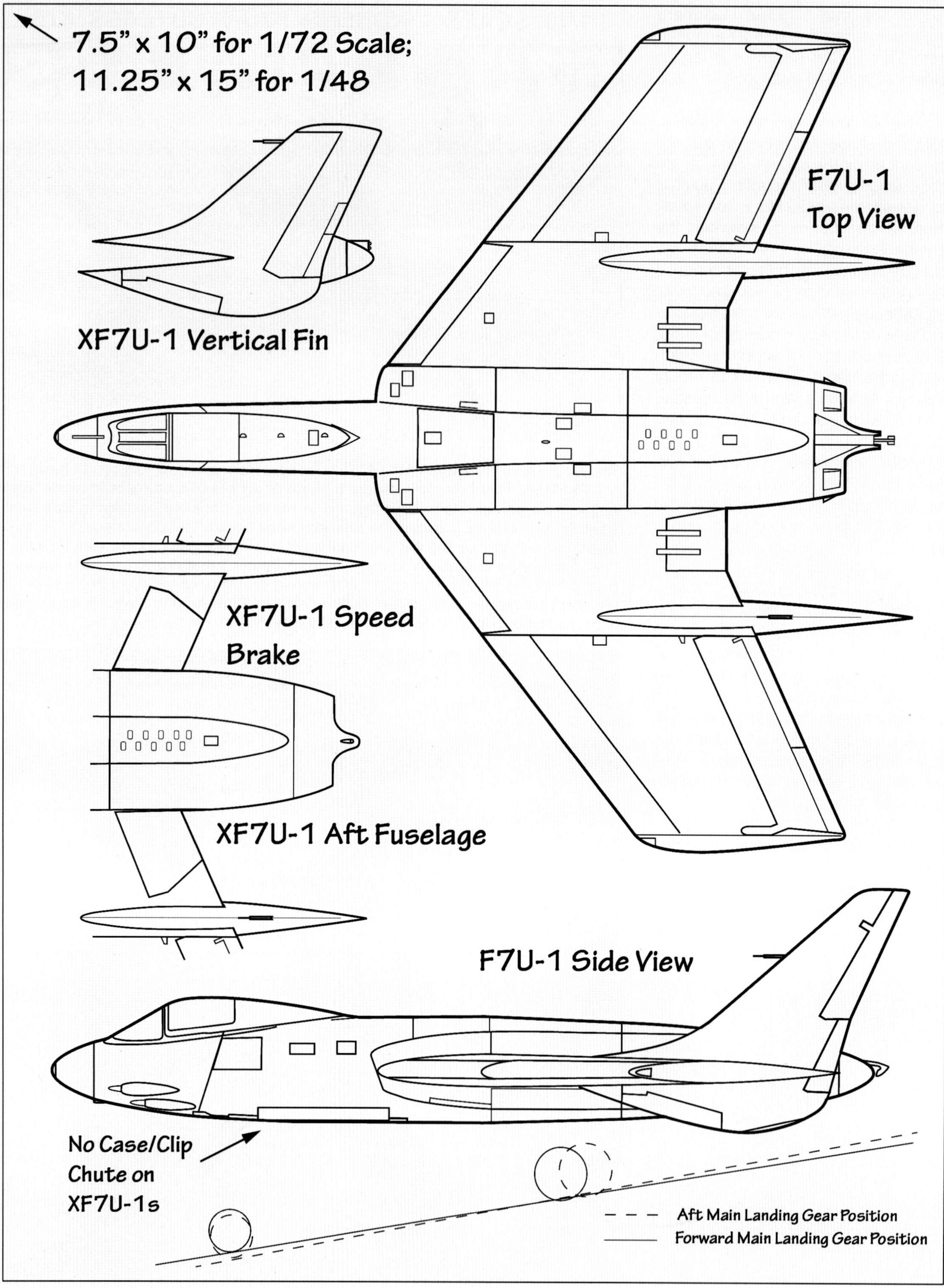
7.5" x 10" for 1/72 Scale;
11.25" x 15" for 1/48
XF7U-1 Vertical Fin
F7U-1
Top View
XF7U-1 Speed
Brake
XF7U-1 Aft Fuselage
F7U-1 Side View
No Case/Clip
Chute on
XF7U-1s
Aft Main Landing Gear Position
Forward Main Landing Gear Position